Machine Learning in Medicine

T0224497

Ton J. Cleophas • Aeilko H. Zwinderman

Machine Learning
in Medicine

by
TON J. CLEOPHAS, MD, PhD, Professor,
Past-President American College of Angiology,
Co-Chair Module Statistics Applied to Clinical Trials,
European Interuniversity College of Pharmaceutical Medicine, Lyon, France,
Department Medicine, Albert Schweitzer Hospital, Dordrecht, Netherlands,

AEILKO H. ZWINDERMAN, MathD, PhD, Professor,
President International Society of Biostatistics,
Co-Chair Module Statistics Applied to Clinical Trials,
European Interuniversity College of Pharmaceutical Medicine, Lyon, France,
Department Biostatistics and Epidemiology, Academic Medical Center, Amsterdam,
Netherlands

With the help from

EUGENE P. CLEOPHAS, MSc, BEng,
HENNY I. CLEOPHAS-ALLERS.

Ton J. Cleophas
European College Pharmaceutical Medicine
Lyon, France

Aeilko H. Zwinderman
Department of Epidemiology
  and Biostatistics
Academic Medical Center
Amsterdam
Netherlands

Please note that additional material for this book can be downloaded from
http://extras.springer.com

ISBN 978-94-007-9363-7   ISBN 978-94-007-5824-7 (eBook)
DOI 10.1007/978-94-007-5824-7
Springer Dordrecht Heidelberg New York London

Preface

Machine learning is a novel discipline concerned with the analysis of large and multiple variables data. It involves computationally intensive methods, like factor analysis, cluster analysis, and discriminant analysis. It is currently mainly the domain of computer scientists, and is already commonly used in social sciences, marketing research, operational research and applied sciences. It is virtually unused in clinical research. This is probably due to the traditional belief of clinicians in clinical trials where multiple variables are equally balanced by the randomization process and are not further taken into account. In contrast, modern computer data files often involve hundreds of variables like genes and other laboratory values, and computationally intensive methods are required. This book was written as a hand-hold presentation accessible to clinicians, and as a must-read publication for those new to the methods.

Some 20 years ago serious statistical analyses were conducted by specialist statisticians. Nowadays there is ready access for professionals without a mathematical background to statistical computing using personal computers or laptops. At this time we witness a second revolution in data-analysis. Computationally intensive methods have been made available in user-friendly statistical software like SPSS software (cluster and discriminant analysis since 2000, partial correlations analysis since 2005, neural networks algorithms since 2006). Large and multiple variables data, although so far mainly the domain of computer scientists, are increasingly accessible for professionals without a mathematical background. It is the authors' experience as master class professors, that students are eager to master adequate command of statistical software. For their benefit, most of the chapters include all of the steps of the novel methods from logging in to the final results using SPSS statistical software. Also for their benefit, SPSS data files of the examples used in the various chapters are available at extras.springer.com.

The authors have given special efforts for all chapters to have their own introduction, discussion, and references sections. They can, therefore, be studied separately and without the need to read the previous chapters first. In addition to the analysis steps of the novel methods explained from data examples, also background information and clinical relevance information of the novel methods is given, and this is done in an explanatory rather than mathematical manner.

We should add that the authors are well-qualified in their field. Professor Zwinderman is president of the International Society of Biostatistics (2012–2015) and Professor Cleophas is past-president of the American College of Angiology (2000–2012). From their expertise they should be able to make adequate selections of modern methods for clinical data analysis for the benefit of physicians, students, and investigators. The authors have been working and publishing together for over 10 years, and their research of statistical methodology can be characterized as a continued effort to demonstrate that statistics is not mathematics but rather a discipline at the interface of biology and mathematics. The authors are not ware of any other work published so far that is comparable with the current work, and, therefore, believe that it does fill a need.

Contents

Chapter 1
Introduction to Machine Learning

1 Summary

1.1 Background

Traditional statistical tests are unable to handle large numbers of variables. The simplest method to reduce large numbers of variables is the use of add-up scores. But add-up scores do not account the relative importance of the separate variables, their interactions and differences in units. Machine learning can be defined as knowledge for making predictions as obtained from processing training data through a computer. If data sets involve multiple variables, data analyses will be complex, and modern computationally intensive methods will have to be applied for analysis.

1.2 Objective and Methods

The current book, using real data examples as well as simulated data, reviews important methods relevant for health care and research, although little used in the field so far.

1.3 Results and Conclusions

1. One of the first machine learning methods used in health research is logistic regression for health profiling where single combinations of x-variables are used to predict the risk of a medical event in single persons (Chap. 2).
2. A wonderful method for analyzing imperfect data with multiple variables is optimal scaling (Chaps. 3 and 4).

T.J. Cleophas and A.H. Zwinderman, *Machine Learning in Medicine*,
DOI 10.1007/978-94-007-5824-7_1, © Springer Science+Business Media Dordrecht 2013

3. Partial correlations analysis is the best method for removing interaction effects from large clinical data sets (Chap. 5).
4. Mixed linear modeling (1), binary partitioning (2), item response modeling (3), time dependent predictor analysis (4) and autocorrelation (5) are linear or loglinear regression methods suitable for assessing data with respectively repeated measures (1), binary decision trees (2), exponential exposure-response relationships (3), different values at different periods (4) and those with seasonal differences (5), (Chaps. 6, 7, 8, 9, and 10).
5. Clinical data sets with non-linear relationships between exposure and outcome variables require special analysis methods, and can usually also be adequately analyzed with neural networks methods like multi layer perceptron networks, and radial basis functions networks (Chaps. 11, 12, and 13).
6. Clinical data with multiple exposure variables are usually analyzed using analysis of (co-) variance (AN(C)OVA), but this method does not adequately account the relative importance of the variables and their interactions. Factor analysis and hierarchical cluster analysis account for all of these limitations (Chaps. 14 and 15).
7. Data with multiple outcome variables are usually analyzed with multivariate analysis of (co-) variance (MAN(C)OVA). However, this has the same limitations as ANOVA. Partial least squares analysis, discriminant analysis, and canonical regression account all of these limitations (Chaps. 16, 17, and 18).
8. Fuzzy modeling is a method suitable for modeling soft data, like data that are partially true or response patterns that are different at different times (Chap. 19).

2   Introduction

Traditional statistical tests are unable to handle large numbers of variables. The simplest method to reduce large numbers of variables is the use of add-up scores. But add-up scores do not account the relative importance of the separate variables, their interactions and differences in units.

Principal components analysis and partial least square analysis, hierarchical cluster analysis, optimal scaling and canonical regression are modern computationally intensive methods, currently often listed as machine learning methods. This is because the computations they make are far too complex to perform without the help of a computer, and because they turn imputed information into knowledge, which is in human terms a kind of learning process.

An additional advantage is that the novel methods are able to account all of the limitations of the traditional methods. Although widely used in the fields of behavioral sciences, social sciences, marketing, operational research and applied sciences, they are virtually unused in medicine. This is a pity given the omnipresence of large numbers of variables in this field of research. However, this is probably just a matter of time, now that the methods are increasingly available in SPSS statistical software and many other packages.

We will start with logistic regression for health profiling where single combinations of x-variables are used to predict the risk of a medical event in single persons (Chap. 2). Then in the Chaps.3 and 4 optimal scaling, a wonderful method for analyzing imperfect data, and in Chap. 5 partial correlations, the best method for removing interaction effects, will be reviewed. Mixed linear modeling (1), binary partitioning (2), item response modeling (3), time dependent predictor analysis (4) and autocorrelation (5) are linear or loglinear regression methods suitable for assessing data with respectively repeated measures (1), binary decision trees (2), exponential exposure-response relationships (3), different values at different periods (4) and those with seasonal differences(5), (Chaps. 6, 7, 8, 9, and 10). Methods for data with non-linear relationships between exposure and outcome variables will be reviewed in the Chaps. 11, 12, and 13, entitled respectively non-linear relationships, artificial intelligence using multilayer perceptron, and artificial intelligence using radial basis functions. Data with multiple exposure variables are usually analyzed using analysis of (co-) variance (AN(C)OVA), but this method does not adequately account the relative importance of the variables and their inter-actions. Also, it rapidly looses power with large numbers of variables relative to the numbers of observations and with strong correlations between the dependent variables. And with large numbers of variables the design matrix may cause integer overflow, too many levels of components, numerical problems with higher order interactions, and commands may not be executed. Factor analysis (Chap. 14) and hierarchical cluster analysis (Chap. 15) account for all of these limitations. Data with multiple outcome variables are usually analyzed with multivariate analysis of (co-) variance (MAN(C)OVA). However, this has the same limitations as ANOVA, and, in addition, with a positive correlation between the outcome variables it is often powerless. The Chaps. 16, 17, and 18 review methods that account for these limitations. These methods are respectively partial least squares analysis, discriminant analysis, and canonical regression. Finally Chap. 19 explains fuzzy modeling as a method for modeling soft data, like data that are partially true or response patterns that are different at different times.

A nice thing about the novel methodologies, thus, is that, unlike the traditional methods like ANOVA and MANOVA, they not only can handle large data files with numerous exposure and outcome variables, but also can do it in a relatively unbiased way.

The current book serves as an introduction to machine learning methods in clinical research, and was written as a hand-hold presentation accessible to clinicians, and as a must-read publication for those new to the methods. It is the authors' experience, as master class professors, that students are eager to master adequate command of statistical software. For their benefit all of the steps of the novel methods from logging in to the final result using SPSS statistical software will be given in most of the chapters. We will end up this initial chapter with some machine learning terminology.

3   Machine Learning Terminology

3.1   Artificial Intelligence

Engineering method that simulates the structures and operating principles of the human brain.

3.2   Bootstraps

Machine learning methods are computationally intensive. Computers make use of bootstraps, otherwise called "random sampling from the data with replacement", in order to facilitate the calculations. Bootstraps is a Monte Carlo method.

3.3   Canonical Regression

Multivariate method. ANOVA / ANCOVA (analysis of (co)variance) and MANOVA / MANCOVA (multivariate analysis of (co)variance) are the standard methods for the analysis of data with respectively multiple independent and dependent variables. A problem with these methods is, that they rapidly lose statistical power with increasing numbers of variables, and that computer commands may not be executed due to numerical problems with higher order calculations among components. Also, clinically, we are often more interested in the combined effects of the clusters of variables than in the separate effects of the different variables. As a simple solution composite variables can be used as add-up sums of separate variables, but add-up sums do not account the relative importance of the separate variables, their interactions, and differences in units. Canonical analysis can account all of that, and, unlike MANCOVA, gives, in addition to test statistics of the separate variables, overall test statistics of entire sets of variables.

3.4   Components

The term components is often used to indicate the factors in a factor analysis, e.g., in rotated component matrix and in principle component analysis.

3.5   Cronbach's alpha

$$alpha = \frac{k}{(k-1)} \cdot \left(1 - \sum \frac{s^2_i}{s^2_T}\right)$$

K = number of original variables
s^2_i = variance of i-th original variable
s^2_T = variance of total score of the factor obtained by summing up all of the original
   variables

3.6 Cross-Validation

Splitting the data into a k-fold scale and comparing it with a k-1 fold scale. Assessment of test-retest reliability of the factors in factor analysis (see internal consistency between the original variables contributing to a factor in factor analysis).

3.7 Data Dimension Reduction

Factor analysis term used to describe what it does with the data.

3.8 Data Mining

A field at the intersection of computer science and statistics, It attempts to discover patterns in large data sets.

3.9 Discretization

Converting continuous variables into discretized values in a regression model.

3.10 Discriminant Analysis

Multivariate method. It is largely identical to factor analysis but goes one step further. It includes a grouping predictor variable in the statistical model, e.g. treatment modality. The scientific question "is the treatment modality a significant predictor of a clinical improvement" is, subsequently, assessed by the question "is the outcome clinical improvement a significant predictor of the odds of having had a particular treatment." This reasoning may seem incorrect, using an outcome for making predictions, but, mathematically, it is no problem. It is just a matter of linear cause-effect relationships, but just the other way around, and it works very conveniently with "messy" outcome variables like in the example given.

3.11 Eigenvectors

Eigenvectors is a term often used with factor analysis. The R-values of the original variables versus novel factors are the eigenvalues of the original variables, their place in a graph the eigenvectors. The scree plot compares the relative importance of the novel factors, and that of the original variables using their eigenvector values.

3.12 Elastic Net Regression

Shrinking procedure similar to lasso, but made suitable for larger numbers of predictors.

3.13 Factor Analysis

Two or three unmeasured factors are identified to explain a much larger number of measured variables.

3.14 Factor Analysis Theory

ALAT (alanine aminotransferase), ASAT (aspartate aminotransferase) and gammaGT (gamma glutamyl tranferase) are a cluster of variables telling us something about a patient's liver function, while ureum, creatinine and creatininine clearance tell us something about the same patient's renal function. In order to make morbidity/ mortality predictions from such variables, often, multiple regression is used. Multiple regression is appropriate, only, if the variables do not correlate too much with one another, and a very strong correlation, otherwise called collinearity or multicollinearity, will be observed within the above two clusters of variables. This means, that the variables cannot be used simultaneously in a regression model, and, that an alternative method has to be used. With factor analysis all of the variables are replaced with a limited number of novel variables, that have the largest possible correlation coefficients with all of the original variables. It is a multivariate technique, somewhat similar to MANOVA (multivariate analysis of variance), with the novel variables, otherwise called the factors, as outcomes, and the original variables, as predictors. However, it is less affected by collinearity, because the y- and x-axes are used to present the novel factors in an orthogonal way, and it can be shown that with an orthogonal relationship between two variables the magnitude of their covariance is zero and has thus not to be taken into account. Factor analysis constructs latent predictor variables from manifest predictor variables, and in this way it can be considered univariate method, but mathematically it is a multivariate method, because multiple rather than single latent variables are constructed from the predictor data available.

3.15   Factor Loadings

The factor loadings are the correlation coefficients between the original variables and the estimated novel variable, the latent factor, adjusted for all of the original variables, and adjusted for eventual differences in units.

3.16   Fuzzy Memberships

The universal spaces are divided into equally sized parts called membership functions.

3.17   Fuzzy Modeling

A method for modeling soft data, like data that are partially true or response patterns that are different at different times.

3.18   Fuzzy Plots

Graphs summarizing the fuzzy memberships of (for example) the imput values.

3.19   Generalization

Ability of a machine learning algorithm to perform accurately on future data.

3.20   Hierarchical Cluster Analysis

It is based on the concept that cases (patients) with closely connected characteristics might also be more related in other fields like drug efficacy. With large data it is a computationally intensive method, and today commonly classified as one of the methods of explorative data mining. It may be more appropriate for drug efficacy analyses than other machine learning methods, like factor analysis, because the patients themselves rather than some new characteristics are used as dependent variables.

3.21   Internal Consistency Between the Original Variables Contributing to a Factor in Factor Analysis

A strong correlation between the answers given to questions within one factor is required: all of the questions should, approximately, predict one and the same thing. The level of correlation is expressed as Cronbach's alpha: 0 means poor, 1 perfect relationship. The test-retest reliability of the original variables should be assessed with one variable missing: all of the data files with one missing variable should produce at least for 80% the same result as that of the non-missing data file (alphas > 80%).

3.22   Iterations

Complex mathematical models are often laborious, so that even modern computers have difficulty to process them. Software packages currently make use of a technique called iterations: five or more calculations are estimated and the one with the best fit is chosen.

3.23   Lasso Regression

Shrinking procedure slightly different from ridge regression, because it shrinks the smallest b-values to 0.

3.24   Latent Factors

The term latent factors is often used to indicate the factors in a factor analysis. They are called latent, because they are not directly measured but rather derived from the original variables.

3.25   Learning

This term would largely fit the term "fitting" in statistics.

3.26   Learning Sample

Previously observed outcome data which are used by a neural network to learn to predict future outcome data as close to the observed values as possible.

3.27   Linguistic Membership Names

Each fuzzy membership is given a name, otherwise called linguistic term.

3.28   Linguistic Rules

The relationships between the fuzzy memberships of the imput data and those of the output data.

3.29   Logistic Regression

Very similar to linear regression. However, instead of linear regression where both dependent and independent variable are continuous, logistic regression has a binary dependent variable (being a responder or non-responder), which is measured as the log odds of being a responder.

3.30   Machine Learning

Knowledge for making predictions, obtained from processing training data through a computer. Particularly modern computationally intensive methods are increasingly used for the purpose.

3.31   Monte Carlo Methods

Iterative testing in order to find the best fit solution for a statistical problem.

3.32   Multicollinearity or Collinearity

There should not be a strong correlation between different original variable values in a conventional linear regression. Correlation coefficient $(R) > 0.80$ means the presence of multicollinearity and, thus, of a flawed multiple regression analysis.

3.33   Multidimensional Modeling

An y- and x-axis are used to represent two factors. If a third factor existed within a data file, it could be represented by a third axis, a z-axis creating a 3-d graph. Also additional factors can be added to the model, but they cannot be presented in a 2- or 3-d drawing, but, just like with multiple regression modeling, the software programs have no problem with multidimensional calculations similar to the above 2-d calculations.

3.34   Multilayer Perceptron Model

Neural network consistent of multiple layers of artificial neurons that after having received a signal beyond some threshold propagates it forward to the next later.

3.35   Multivariate Machine Learning Methods

The methods that always include multiple outcome variables. They include discriminant analysis, canonical regression, and partial least squares.

3.36   Multivariate Method

Statistical analysis method for data with multiple outcome variables.

3.37   Network

This term would largely fit the term "model" in statistics.

3.38   Neural Network

Distribution-free method for data modeling based on layers of artificial neurons that transmit imputed information.

3.39   Optimal Scaling

The problem with linear regression is that consecutive levels of the variables are assumed to be equal, but in practice this is virtually never true. Optimal scaling is a method designed to maximize the relationship between a predictor and an outcome variable by adjusting their scales. It makes use of discretization and regularization methods (see there).

3.40   Overdispersion, Otherwise Called Overfitting

The phenomenon that the spread in the data is wider than compatible with Gaussian modeling. This phenomenon is particularly common with discretization of continuous variables.

3.41   Partial Correlation Analysis

With a significant interaction between independent variables an overall data analysis is meaningless. Instead the study could be repeated with the interacting variables held constant. The next best method is partial correlation analysis, where the effects of interaction is removed by artificially holding the interacting variables constant.

3.42   Partial Least Squares

Multivariate method. Like factor analysis it identifies latent variables. However, partial least squares analysis does not simultaneously use all of the predictor variables available, but rather uses an a priori clustered set of predictor variables of 4 or 5 to calculate a new latent variable. Unlike factor analysis which does not consider response variables at all, partial least square analysis does take response variables into account and therefore often leads to a better fit of the response variable. Correlation coefficients are produced from multivariate linear regression rather than fitted correlation coefficients along the x and y-axes.

3.43   Pearson's Correlation Coefficient (R)

$$R = \frac{\sum (x - \bar{x})(y - \bar{y})}{\sqrt{\sum (x - \bar{x})^2 \sum (y - \bar{y})^2}}$$

R is a measure for the strength of association between two variables. The stronger the association, the better one variable predicts the other. It varies between −1 and +1, zero means no correlation at all, −1 means 100% negative correlation, +1 100% positive correlation.

3.44   Principal Components Analysis

Synonym of factor analysis.

3.45   Radial Basis Functions

Symmetric functions around the origin, the equations of Gaussian curves are radial basis functions.

3.46   Radial Basis Function Network

Neural network, that, unlike the multilayer perceptron network, uses Gaussian instead of sigmoidal activation functions for transmission of signals.

3.47   Regularization

Correcting discretized variables for overfitting, otherwise called overdispersion.

3.48   Ridge Regression

Important method for shrinking b-values for the purpose of adjusting overdispersion.

3.49   Splines

Cut pieces of a non-linear graph, originally thin wooden strips for modeling cars and airplanes.

3.50   Supervised Learning

Machine learning using data that include both input and output data (exposure and outcome data).

3.51   Training Data

The output data of a supervised learning data set.

3.52   Triangular Fuzzy Sets

A common way of drawing the membership function with on the x-axis the imput values, on the y-axis the membership grade for each imput value.

3.53   Universal Space

A term often used with fuzzy modeling: the defined range of imput values, and defined range of output values.

3.54   Unsupervised Learning

Machine learning using data that includes only input data (exposure data).

3.55   Varimax Rotation

It can be demonstrated in a "2 factor" factor analysis, that, by slightly rotating both x and y-axes, the model can be fitted even better. When the y- and x-axes are rotated simultaneously, the two novel factors are assumed to be 100% independent of one another, and this rotation method is called varimax rotation. Independence needs not be true, and, if not true, the y-axis and x-axis can, alternatively, be rotated separately in order to find the best fit model for the data given.

3.56   Weights

This term would largely fit the term "parameters" in statistics.

The above terms are just a few, and many more novel terms are being used by computer scientists involved in machine learning. Some more terms will be dealt with in the subsequent chapters. Also in the index a better overview of them will be given.

4   Discussion

Sometimes machine learning is discussed as a discipline conflicting with statistics [1]. Indeed, there are differences, particularly, in terminology as reviewed above. Generally machine learning is performed by computer scientists with a background of psychology, biology, economics, not by statisticians. In contrast, statisticians commonly have a mathematical background. Computer science is a very modern discipline with "cooler" terminologies, better jobs and better income perspectives than statistics. Computer science, usually, involves larger and more complex data sets, and, continually, makes use of prediction modeling rather than null hypothesis testing. However, because of their relatively weak mathematical background, computer scientists are sometimes not well aware of the limitations of the models they use.

5   Conclusions

1. Machine learning can be defined as knowledge for making predictions as obtained from processing training data through a computer. Because data sets often involve multiple variables, data analyses are complex and modern computationally intensive methods have to be used for the purpose. The current book reviews important methods relevant for health care and research, although little used in the field so far.
2. One of the first machine learning methods used in health research is logistic regression for health profiling where single combinations of x-variables are used to predict the risk of a medical event in single persons (Chap. 2).
3. A wonderful method for analyzing imperfect data with multiple variables is optimal scaling (Chaps. 3 and 4).
4. Partial correlations analysis is the best method for removing interaction effects from large clinical datasets (Chap. 5).
5. Mixed linear modeling (1), binary partitioning (2), item response modeling (3), time dependent predictor analysis (4) and autocorrelation (5) are linear or loglinear regression methods suitable for assessing data with respectively repeated measures (1), binary decision trees (2), exponential exposure-response relationships (3), different values at different periods (4) and those with seasonal differences (5), (Chaps. 6, 7, 8, 9, and 10).
6. Clinical data sets with non-linear relationships between exposure and outcome variables require special analysis methods, and can usually also be adequately analyzed with neural networks methods like multilayer perceptron networks, and radial basis functions networks (Chaps. 11, 12, and 13).
7. Clinical data with multiple exposure variables are usually analyzed using analysis of (co-) variance (AN(C)OVA), but this method does not adequately account the relative importance of the variables and their interactions. Factor analysis and hierarchical cluster analysis account for all of these limitations (Chaps. 14 and 15).

8. Data with multiple outcome variables are usually analyzed with multivariate analysis of (co-) variance (MAN(C)OVA). However, this has the same limitations as ANOVA. Partial least squares analysis, discriminant analysis, and canonical regression account all of these limitations (Chaps. 16, 17, and 18).
9. Fuzzy modeling is a method suitable for modeling soft data, like data that are partially true or response patterns that are different at different times (Chap. 19).

Reference

1. O'Connor B (2012) Statistics versus machine learning, fight. http://brenocon.com/blog/2008/12. Accessed 25 Aug 2012

Chapter 2
Logistic Regression for Health Profiling

1 Summary

1.1 Background

Logistic regression can be used for predicting the probability of an event in subjects at risk.

1.2 Methods and Results

It uses log linear models of the kind of the one underneath (ln = natural logarithm, a = intercept, b = regression coefficient, x = predictor variable):

$$\text{"ln odds infarct} = a + b_1 x_1 + b_2 x_2 + b_3 x_3 + \cdots \text{."}$$

A real data example was used of 1,000 subjects of different ages followed for 10 years for myocardial infarctions. Using the data, an exact risk could be calculated for individual future subjects.

1.3 Conclusions

1. The methodology is currently an important way to determine, with limited health care sources, what individuals are at low risk and will, thus, be:
   (1) operated.
   (2) given expensive medications.
   (3) given the assignment to be treated or not.
   (4) given the "do not resuscitate sticker".
   (5) etc.

T.J. Cleophas and A.H. Zwinderman, *Machine Learning in Medicine*,
DOI 10.1007/978-94-007-5824-7_2, © Springer Science+Business Media Dordrecht 2013

Fig. 2.1 In a group of
multiple ages the numbers
of patients at risk of
infarction is given by the
dotted line

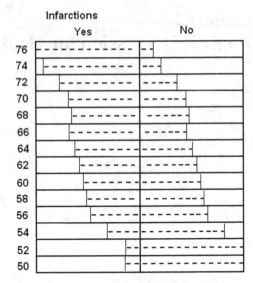

2. We must take into account that some of the predictor variables may be heavily
   correlated with one another, and the results may, therefore, be inflated.
3. Also, the calculated risks may be true for subgroups, but for individuals less so,
   because of the random error.

2   Introduction

Logistic regression can be used for predicting the probability of an event. For
example, the odds of an infarction is given by the equation

$$\text{odds infarct in a group} = \frac{\text{number of patients with infarct}}{\text{number of patients without}}$$

The odds of an infarction in a group is correlated with age, the older the patient
the larger the odds

According to Fig. 2.1 the odds of infarction is correlated with age, but we may
ask how?

According to Fig. 2.2 the relationship is not linear, but after transformation of
the odds values on the y-axis into log odds values the relationship is suddenly
linear.

We will, therefore, transform the linear equation

$$y = a + bx$$

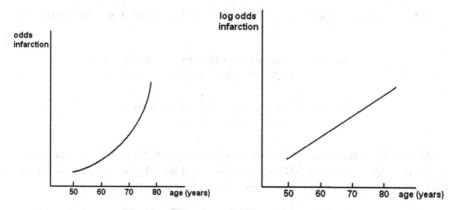

Fig. 2.2 Relationships between the odds of infarction and age

into a log linear equation (ln = natural logarithm)

$$\ln \text{ odds} = a + b \text{ x } (x = \text{age}).$$

3   Real Data Example

Our group consists of 1,000 subjects of different ages that have been observed for 10 years for myocardial infarctions. Using SPSS statistical software, we command binary logistic regression

> dependent variable infarction yes/no $(0 / 1)$
> independent variable age

The program produces a regression equation:

$$\ln \text{ odds} = \ln \frac{\text{pts with infarctions}}{\text{pts without}} = a + bx$$

a = −9.2
b = 0.1 (SE = 0.04; p < 0.05)

The age is, thus, a significant determinant of odds infarction (which can be used as surrogate for risk of infarction).

Then, we can use the equation to predict the odds of infarction from a patient's age:

$$\text{Ln odds 55 years} = -9.2 + 0.1 . 55 = -4.82265$$
$$\text{odds} = 0.008 = 8 / 1,000$$

$$\text{Ln odds 75 years} = -9.2 + 0.1 . 75 = -1.3635$$
$$\text{odds} = 0.256 = 256 / 1,000$$

Odds of infarction can, of course, more reliably be predicted from multiple x-variables. As an example, 10,000 pts are followed for 10 years, while infarctions and baseline-characteristics are registered during that period.

dependent variable	infarction yes/no
independent variables (predictors)	gender
	age
	Bmi (body mass index)
	systolic blood pressure
	cholesterol
	heart rate
	diabetes
	antihypertensives
	previous heart infarct
	smoker

The data are entered in SPSS, and it produces b-values (predictors of infarctions)

	b-values	p-value
1. Gender	0.6583	<0.05
2. Age	0.1044	"
3. Bmi	−0.0405	"
4. Systolic blood pressure	0.0070	"
5. Cholesterol	0.0008	"
6. Heart rate	0.0053	"
7. Diabetes	1.2509	<0.10
8. Antihypertensives	0.3175	<0.050
9. Previous heart infarct	0.8659	<0.10
10.Smoker	0.0234	<0.05
a-value	−9.1935	"

It is decided to exclude predictors that have a p-value >0.10. The underneath regression equation is used

$$\text{"ln odds infarct} = a + b_1 x_1 + b_2 x_2 + b_3 x_3 + ..."$$

to calculate the best predictable y-value from every single combination of x-values.

For instance, for a subject with the following characteristics (= predictor variables)

- Male (x_1)
- 55 years of age (x_2)
- cholesterol 6.4 mmol/l (x_3)
- systolic blood pressure 165 mmHg (x_4)
- antihypertensives (x_5)
- dm (x_6)
- 15 cigarettes/day (x_7)
- heart rate 85 beats/min (x_8)
- Bmi 28.7 (x_9)
- smoker (x_{10})

the calculated odds of having an infarction in the next 10 years is the following:

	b-values	x-values	
Gender	0.6583 .	1 (0 or 1)	= 0.6583
Age	0.1044 .	55	= 5.742
BMI	−0.0405 .	28.7	= ..
Blood pressure	0.0070 .	165	=
Cholesterol	0.0008 .	6.4	=
Heart rate	0.0053 .	85	=
Diabetes	1.2509 .	1	=
Antihypertensives	0.3175 .	1	=
Previous heart inf	0.8659 .	0	=
Smoker	0.0234 .	15	=
a-value			= −9.1935 +
		Ln odds infarct	= −0.5522
		odds infarct	= 0.58 = 58/100

The odds is often interpreted as risk. However, the true risk is a bit smaller than the odds, and can be found by the equation

$$\text{risk event} = 1/(1+1/\text{odds})$$

If odds of infarction = 0.58, then the true risk of infarction = 0.37.

4 Discussion

The above methodology is currently an important way to determine, with limited health care sources, what individuals will be:

1. operated.
2. given expensive medications.

Table 2.1 Examples of predictive models where multiple logistic regression has been applied

Dependent variable (odds of event)	Independent variables (predictors)
1. TIMI risk score [1]	
Odds of infarction	Age, comorbidity, comedication, riskfactors
2. Car producer (Strategic management research) [2]	
Odds of successful car	Cost, size, horse power, ancillary properties
3. Item response modeling (Rasch models for computer adapted tests) [3]	
Odds of correct answer to three questions of different difficulty	Correct answer to three previous questions

3. given the assignment to be treated or not.
4. given the "do not resuscitate sticker".
5. etc.

We need a large data base to obtain accurate b-values. This logistic model for turning the information from predicting variables into probability of events in individual subjects is being widely used in medicine, and was, for example, the basis for the TIMI (Thrombolysis In Myocardial Infarction) prognostication risk score. However, not only in medicine, also in strategic management research, psychological tests like computer adapted tests, and many more fields it is increasingly observed (Table 2.1). With linear regression it is common to provide a measure of how well the model fits the data, and the squared correlation coefficient r^2 is mostly applied for that purpose. Unfortunately, no direct equivalent to r^2 exists for logistic, otherwise called loglinear, models. However, pseudo-R2 or R2-like measures for estimating the strength of association between predictor and event have been developed.

Logistic regression is increasingly used for predicting the risk of events like cardiovascular events, diagnosis of cancer, deaths etc. Limitations of this approach have to be taken into account. It uses observational data and may, consequently, give rise to serious misinterpretations of the data:

1. The assumption that baseline characteristics are independent of treatment efficacies may be wrong.
2. Sensitivity of testing is jeopardized if the models do not fit the data well enough.
3. Relevant clinical phenomena like unexpected toxicity effects and complete remissions can go unobserved.
4. The inclusion of multiple variables in regression models raises the risk of clinically unrealistic results.

As another example, a cohort of postmenopausal women is assessed for the risk of endometrial cancer. The main question is: what are the determinants of endometrial cancer in a category of females. The following logistic model is used:

y-variable = ln odds endometrial cancer
x_1 = estrogene consumption short term

x_2 = estrogene consumption long term
x_3 = low fertility index
x_4 = obesity
x_5 = hypertension
x_6 = early menopause

lnodds endometrial cancer = $a + b_1$ estrogene data + b_2 .... + b_6 early menopause data
The odds ratios for different x-variables are defined, e.g., for:

x_1 = chance cancer in consumers of estrogene/non-consumers
x_3 = chance cancer in patients with low fertility/their counterparts
x_4 = chance cancer in obese patients/their counterparts etc.

risk factors	regression coefficient(b)	standard error	p-value	odds ratio (e^b)
1. estrogenes short	1.37	0.24	<0.0001	3.9
2. estrogenes long	2.60	0.25	<0.0001	13.5
3. low fertility	0.81	0.21	0.0001	2.2
4. obesity	0.50	0.25	0.04	1.6
5. hypertension	0.42	0.21	0.05	1.5
6. early menopause	0.53	0.53	ns	1.7

The data were entered in the software program, which provided us with the best fit b-values. The model not only showed a greatly increased risk of cancer in several categories, but also allowed to consider that the chance of cancer if patients consume estrogens, suffer from low fertility, obesity, and hypertension might have an increased risk as large as $= e^{b2+b3+b4+b5} = 75.9 = 76$ fold. This huge chance is, of course, clinically unrealistic! We must take into account that some of these variables must be heavily correlated with one another, and the results are, therefore, largely inflated. In conclusion, logistic regression is an adequate tool for exploratory research, the conclusions of which must be interpreted with caution, although they often provide scientifically highly interesting questions. This should be kept in mind when using it for health profiling in individuals. Also the calculated risk may be true for subgroups, but for individuals less so, because of the random error.

5 Conclusions

Logistic regression can be used for predicting the probability of an event. It uses log linear models of the kind of the one underneath:

$$\text{"ln odds infarct} = a + b_1 x_1 + b_2 x_2 + b_3 x_3 + \cdots \text{."}$$

The methodology is currently an important way to determine, with limited health care sources, what individuals will be:

1. operated.
2. given expensive medications.
3. given the assignment to be treated or not.
4. given the "do not resuscitate sticker".
5. etc.

We must take into account that some of these variables must be heavily correlated with one another, and the results are, therefore, largely inflated. Also the calculated risks may be true for subgroups, but for individuals less so, because of the random error.

References

1. Antman EM, Cohen M, Bernink P, McGabe CH, Horacek T, Papuches G, Mautner B, Corbalan R, Radley D, Braunwald E (2000) The TIMI risk score for unstable angina pectors, a method for prognostication and therapeutic decision making. J Am Med Assoc 284:835–842
2. Hoetner G (2007) The use of logit and probit models in strategic management research. Strateg Manag J 28:331–343
3. Rudner LM Computer adaptive testing. http://edres.org/scripts/cat/catdemo.htm. Accessed 18 Dec 2012

Chapter 3
Optimal Scaling: Discretization

1 Summary

1.1 Background

In clinical trials the research question is often measured with multiple variables, and multiple regression is commonly used for analysis. The problem with multiple regression is that consecutive levels of the variables are assumed to be equal, while in practice this is virtually never true. Optimal scaling is a method designed to maximize the relationship between a predictor and an outcome variable by adjusting their scales.

1.2 Objective and Methods

A simulated example of a drug efficacy trial including 27 variables was used to compare the performance of optimal scaling with that of traditional multiple linear regression. The SPSS module Optimal Scaling was used.

1.3 Results

The two methods produced similarly sized results with 7 versus 6 p-values < 0.10 and 3 versus 4 p-values < 0.010 respectively.

T.J. Cleophas and A.H. Zwinderman, *Machine Learning in Medicine*,
DOI 10.1007/978-94-007-5824-7_3, © Springer Science+Business Media Dordrecht 2013

1.4   Conclusions

1. Optimal scaling using discretization is a method for analyzing clinical trials where the consecutive levels of the variables are inequal.
2. In order to fully benefit from optimal scaling a regularization procedure for the purpose of correcting overdispersion is desirable.

2   Introduction

In clinical trials the research question is often measured with multiple variables. For example, the expressions of a number of genes can be used to predict the efficacy of cytostatic treatment [1, 2], repeated measurements can be used in randomized longitudinal trials [3, 4], and multi-item personal scores can be used for the evaluation of antidepressants [5]. Many more examples can be given. Multiple linear regression analysis is often used for analyzing the effect of predictors (x-axis variables) on outcome variables (y-axis variables). A problem with linear regression is that consecutive levels of the predictor variables are assumed to be equal, while in practice this is virtually never true. Figure 3.1 gives an example of a continuous predictor variable scored on a scale of 0–10. Patients with the predictor values 0, 1, 5, 9 and 10 are missing. Instead of a scale of integers between 0 and 10, other scales are possible, e.g. a scale of two or four parts (Table 3.1). Any scale used is, of course, arbitrary and can be replaced with another one. In the example of Fig. 3.1 the following scales are used.

Scale 1: 0, 1, 2, 3, 4, 5, 6, 7, 8, 9, 10.
Scale 2: 1, 2 (1 = 0–5 from scale 1; 2 = 5–10).
Scale 3: 1, 2, 3, 4 (1 = 0–2.5 from scale 1; 2 = 2.5–5; 3 = 5–7.5; 4 = 7.5–10).

The Table 3.1 shows that each scale produced a different pattern of results with one result better than the other. With the scales 2 and 3 a gradual improvement of the t-values and p-values is observed. Optimal scaling is a method designed to maximize the relationship between a predictor and an outcome variable. It is a computationally intensive method invented by Albert Gifi, professor of statistics at UCLA, Los Angeles CA, 1990, and uses a mathematical technique called the quadratic approximation to find out of an endless number of scales the very best one for your data [6]. Optimal scaling is widely reviewed in the statistical literature, and is an important branch of machine learning, a discipline concerned with computers to make predictions based on complex empirical data [7–10]. In clinical research it is little used so far. When searching Medline we only found a few genetic studies [11, 12], epidemiological [13, 14] and psychological [15] studies, but very few therapeutic trials [16, 17], despite its pleasant property to improve the p-values of testing and, thus, turn negative results into positive ones.

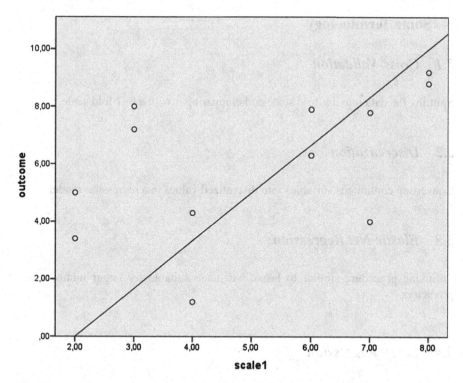

Fig. 3.1 Linear regression analysis. An example of a continuous predictor variable (x-variable) on a scale 0–10. Patients with the predictor values 0, 1, 5, 9 and 10 are missing

Table 3.1 Linear regression analysis of the data from Fig. 3.1 using three different scales

| | Coefficients[a] | | | | |
| | Unstandardized coefficients | | Standardized coefficients | | |
Model	B	Std. Error	Beta	t	Sig.
1 (Constant)	3,351	1,647		2,034	,069
Scale1	,548	,302	,497	1,813	,100
1 (Constant)	2,367	2,032		1,165	,271
Scale2	,497	,257	,521	1,932	,082
1 (Constant)	2,217	1,647		1,346	,208
Scale3	,620	,246	,623	2,520	,030

With the scales 2 and 3 a gradual improvement of the t-values and p-values is observed
[a]Dependent variable: outcome

The current chapter uses a simulated example to assess the performance of optimal scaling in a multiple variables model. We hope this chapter will stimulate clinical investigators to start using this optimized analysis method for predictive trials.

3   Some Terminology

3.1   Cross-Validation

Splitting the data into a k-fold scale and comparing it with a k−1 fold scale.

3.2   Discretization

Converting continuous variables into discretized values in a regression model.

3.3   Elastic Net Regression

Shrinking procedure similar to lasso, but made suitable for larger numbers of predictors.

3.4   Lasso Regression

Shrinking procedure slightly different from ridge regression, because it shrinks the smallest b-values to 0.

3.5   Overdispersion, Otherwise Called Overfitting

The phenomenon that the spread in the data is wider than compatible with Gaussian modeling. This phenomenon is particularly common with discretization of continuous variables.

3.6   Monte Carlo Methods

Iterative testing in order to find the best fit solution for a statistical problem.

3.7   Regularization

Correcting discretized variables for overfitting, otherwise called overdispersion.

3.8 Ridge Regression

Important method for shrinking b-values for the purpose of adjusting overdispersion.

3.9 Splines

Cut pieces of a non-linear graph, originally thin wooden strips for modeling cars and airplanes.

4 Some Theory

Optimal scaling makes use of processes like discretization (converting continuous variables into discretized values), and regularization (correcting discretized variables for overfitting, otherwise called overdispersion) [8–11].

In order to transform continuous data into a discrete model, the quadratic approximation is convenient: $fx = fa + f[1]a(x-a)$ where $f[1]a$ is the first derivative of the function fa. The quadratic approximation is based on the principle that the simplest model next to the linear is the quadratic model. Obviously, the magnitude of (any) function can be described by the first derivative of the same function (= slope of the function). This approach is helpful for assessing complex functions like those of standard errors, but also to find the best fit distance (discretization) between some x-value and an x-value close by, called an a-value, which is then used as the best fit scale for the data.

In order to further improve the best fit scale for a variable, SPSS provides the possibility to cut a linear variable into two pieces (splines), combining two linear functions for modeling not entirely linear patterns.

5 Example

A 250 patients' data-file was supposed to include 27 variables consistent of both patients' microarray gene expression levels and their drug efficacy scores. The data file is in the appendix. All variables were standardized by scoring them on 11 points linear scales (0–10). The following genes were highly expressed: the genes 1–4, 16–19, and 24–27. As outcome variable composite scores of the variables 20–23 were used.

For analyses SPSS statistical software [18] was used. Table 3.2 shows the results of a traditional multiple linear regression with the gene expression levels as predictors and the drug efficacy composite score as outcome. The overall r-square value was

Table 3.2 Traditional multiple linear regression with drug efficacy score (a composite score of the variables 20–23) as outcome and 12 gene expression levels as predictor

		Coefficients[a]				
		Unstandardized coefficients		Standardized coefficients		
Model		B	Std. Error	Beta	t	Sig.
1	(Constant)	3,567	1,479		2,411	,017
	Geneone	−,123	,190	−,031	−,649	,517
	Genetwo	,294	,225	,081	1,306	,193
	Genethree	,380	,228	,100	1,662	,098
	Genefour	,052	,197	,012	,264	,792
	Genesixteen	,730	,172	,232	4,241	,000
	Geneseventeen	,809	,198	,216	4,078	,000
	Geneeighteen	,087	,152	,027	,570	,569
	Genenineteen	,596	,154	,196	3,871	,000
	Genetwentyfour	,357	,146	,137	2,443	,015
	Genetwentyfive	,025	,141	,007	,174	,862
	Genetwentysix	,362	,143	,142	2,539	,012
	Genetwentyseven	−,262	,134	−,088	−1,964	,051

[a]Dependent variable: 20–23

0.725. In order to improve the scaling of the linear regression model the Optimal Scaling program of SPSS was used.

> Command: Analyze….Regression….Optimal Scaling….Dependent Variable: Var 28 (Define Scale: mark spline ordinal 2.2)….Independent Variables: Var 1, 2, 3, 4, 16, 17, 18, 19, 24, 25, 26, 27 (all of them Define Scale: mark spline ordinal 2.2)….Discretize: Method Grouping, Number categories 7)….OK.

The Table 3.3 shows the results. There is no intercept anymore and the t-tests have been replaced with F-tests. The optimally scaled model without regularization shows similarly sized effects compared to those of the traditional multiple linear regression model, although the overall r-square was slightly larger with a value of 0.736 versus 0.725, and so was the number of significant p-values <0.010, namely 4 versus 3 respectively. In order, however, to fully benefit from optimal scaling a regularization procedure for the purpose of correcting overdispersion is desirable. It will be addressed in the next chapter.

6   Discussion

Traditional linear regression has been demonstrated not to perform well in case of multiple independent variables. This is because consecutive levels of variables are often inequal. Optimal scaling is a method adequate for such variables. It may however cause power loss due to overdispersion [9, 10].

Table 3.3 Optimal scaling without regularization

Coefficients

| | Standardized coefficients | | | | |
	Beta	Boots trap (1,000) Estimate of Std. error	df	F	Sig.
Geneone	−,092	,110	2	,701	,497
Genetwo	,181	,098	4	3,420	,010
Genethree	−,087	,125	3	,483	,694
Genefour	,114	,065	2	3,029	,051
Genesixteen	,230	,088	4	6,866	,000
Geneseventeen	,273	,119	2	5,272	,006
Geneeighteen	,144	,144	1	,998	,319
Genenineteen	,153	,069	2	4,920	,008
Genetwentyfour	,169	,092	2	3,394	,036
Genetwentyfive	−,023	,095	1	,060	,807
Genetwentysix	,111	,109	2	1,040	,355
Genetwentyseven	−,116	,107	3	1,174	,321

Dependent variable: 20–23

Also, a sharp increase of the t-values of some x-values is often observed if other x-values are removed. This phenomenon is called instable regression coefficients, sometimes referred to as "bouncing betas" [8], and arises when predictors are correlated or when a large number of predictors relative to the number of observations is present. Shrinking the regression coefficients has been demonstrated to be beneficial not only to counterbalance overdispersion but also to reduce this phenomenon of instability [9, 10].

The method of optimal scaling is computationally intensive and sometimes listed as a machine learning method, because it produces predictive models with help of a computer program. Other machine learning methods include factor analysis, partial least squares (pls), canonical regression, item response modeling, neural networks etc [19]. We should emphasize that the optimal scaling method is not a competitor of the others, but rather a different approach. It is particularly beneficial with lack of homogeneity in the scales of the variables. Factor analysis [20] and pls [21] are more suitable if you are interested in the combined effects of subsets of variables, and canonical regression [22] is adequate if you are mostly interested to assess the combined effects of all of your predictor variables.

Limitations of the method should be mentioned. First independence of the scale intervals on the magnitude of the outcome variable is an assumption that may not be appropriate in some cases. Also the spread in the data is sometimes inappropriately amplified with wide scale intervals. Finally, more than a single scale may be optimal [23].

The current chapter shows that in order to fully benefit from optimal scaling a regularization procedure is desirable. Three methods are explained.

7   Conclusion

1. Optimal scaling using discretization is a method for analyzing clinical trials where the consecutive levels of the variables are inequal.
2. In order to fully benefit from optimal scaling a regularization procedure for the purpose of correcting overdispersion is desirable.

8   Appendix: Datafile of 250 Subjects Used as Example

G1	G2	G3	G4	G16	G17	G18	G19	G24	G25	G26	G27	O1	O2	O3	O4
8	8	9	5	7	10	5	6	9	9	6	6	6	7	6	7
9	9	10	9	8	8	7	8	8	9	8	8	8	7	8	7
9	8	8	8	8	9	7	8	9	8	9	9	9	8	8	8
8	9	8	9	6	7	6	4	6	6	5	5	7	7	7	6
10	10	8	10	9	10	10	8	8	9	9	9	8	8	8	7
7	8	8	8	8	7	6	5	7	8	8	7	7	6	6	7
5	5	5	5	5	6	4	5	5	6	6	5	6	5	6	4
9	9	9	9	8	8	8	8	9	8	3	8	8	8	8	8
9	8	9	8	9	8	7	7	7	7	5	8	8	7	6	6
10	10	10	10	10	10	10	10	10	8	8	10	10	10	9	10
2	2	8	5	7	8	8	8	9	3	9	8	7	7	7	6
7	8	8	7	8	6	6	7	8	8	8	7	8	7	8	8
8	9	9	8	10	8	8	7	8	8	9	9	7	7	8	8
7	7	8	8	8	9	10	7	9	4	8	8	9	8	7	7
3	4	3	8	4	4	4	3	4	3	4	4	4	4	3	4
7	8	8	5	8	8	7	6	7	7	8	7	10	8	8	7
8	8	8	8	6	8	5	1	9	7	7	8	7	7	8	6
7	8	8	8	8	9	8	7	10	10	9	8	9	9	9	9
8	4	3	8	3	5	5	3	2	10	1	0	5	3	4	3
8	7	6	10	8	8	7	6	4	4	5	5	7	7	7	5
9	9	10	8	8	9	7	7	8	9	8	9	8	7	8	7
6	6	6	6	4	5	4	5	3	9	3	4	4	5	4	3
8	8	8	7	7	7	8	6	8	7	9	4	6	7	8	9
9	9	10	9	10	10	7	10	10	10	10	10	8	8	8	5
8	7	8	8	9	8	9	8	8	8	8	8	8	8	8	9
8	5	5	4	2	1	1	0	0	1	0	0	3	2	4	5
6	6	6	6	5	6	3	5	4	4	4	5	5	6	3	4
7	8	9	8	8	9	9	6	9	8	8	10	9	8	7	7
8	8	8	7	7	7	7	6	7	8	7	8	7	6	6	6
8	8	8	8	9	8	9	8	9	8	9	9	9	8	7	8
7	7	7	6	7	7	9	7	7	7	7	8	8	6	7	7
9	9	9	9	6	9	8	7	8	8	8	9	8	8	8	8
10	10	10	10	9	9	10	5	10	2	9	9	8	10	8	8
9	8	9	9	8	7	7	8	9	9	9	9	8	5	9	7

(continued)

(continued)

G1	G2	G3	G4	G16	G17	G18	G19	G24	G25	G26	G27	O1	O2	O3	O4
8	9	9	9	8	7	7	6	7	8	8	8	8	7	8	6
3	4	2	5	4	2	2	4	4	4	3	4	6	2	3	2
8	8	9	9	8	8	8	8	8	8	8	8	8	8	8	8
8	6	7	6	7	7	8	6	7	6	5	5	6	7	7	6
10	10	10	10	7	10	10	8	10	10	10	10	10	8	8	7
8	10	9	8	8	8	7	6	7	7	10	8	9	8	8	7
8	8	8	8	8	8	7	8	7	8	8	8	9	9	8	7
5	7	7	8	5	7	7	3	1	6	3	10	5	6	6	5
10	9	9	10	7	9	9	9	9	9	9	8	8	9	7	7
9	7	7	9	3	6	4	2	1	8	2	1	6	6	6	6
8	8	10	8	9	8	7	8	8	7	8	8	9	6	5	7
6	8	8	8	9	10	10	9	10	9	9	10	9	8	5	5
8	8	8	8	10	8	7	10	8	8	7	10	9	7	8	6
6	5	5	6	6	6	4	6	3	5	0	3	7	5	5	3
9	9	9	8	8	9	8	7	6	7	8	10	8	8	8	6
9	10	8	8	9	10	10	9	7	8	9	7	8	8	7	7
8	8	8	9	6	8	7	6	8	9	8	8	7	7	6	5
8	5	6	7	8	8	7	7	4	6	7	6	8	8	7	6
4	1	4	9	0	0	7	0	0	10	0	10	0	0	0	0
5	5	7	5	7	7	8	5	7	7	5	5	7	7	7	7
5	5	6	5	4	4	4	3	3	2	3	3	3	4	3	3
7	9	9	10	5	9	9	9	9	6	7	6	10	7	10	9
10	10	10	10	8	9	9	6	7	8	8	10	7	7	7	6
8	8	8	8	6	9	8	7	7	6	6	2	7	7	7	5
6	6	7	9	8	8	7	6	1	9	0	4	6	7	7	6
6	7	7	7	6	5	5	5	5	7	3	5	7	6	6	8
9	9	9	9	8	8	9	6	8	7	6	10	8	7	7	8
7	7	7	7	6	8	8	6	7	7	7	8	6	6	5	10
9	7	8	9	8	10	8	9	8	9	7	9	7	7	8	3
8	9	9	8	7	8	7	8	8	6	7	8	7	8	7	6
8	8	8	8	6	8	8	5	8	9	8	7	7	7	6	5
7	7	7	7	4	5	6	6	3	6	7	7	1	5	6	5
9	10	9	9	8	9	8	8	9	8	9	9	8	7	8	8
8	9	9	8	8	8	8	7	8	7	8	9	6	6	5	6
7	8	8	8	6	7	7	6	8	5	7	7	7	6	7	5
4	2	2	6	5	5	4	4	6	4	3	2	4	6	7	2
5	5	7	5	5	5	5	2	2	9	5	5	4	5	5	4
9	9	10	9	7	8	7	8	8	9	8	8	8	8	6	9
8	8	8	8	7	7	7	9	8	9	7	8	7	7	5	6
8	8	9	8	8	9	5	9	8	5	7	6	8	8	8	6
9	9	9	9	6	8	8	4	7	5	6	6	7	7	8	8
9	8	8	8	7	9	9	9	10	10	10	10	10	9	7	10
9	9	9	8	8	8	8	7	7	7	7	7	8	8	8	7
8	5	7	9	2	8	8	2	9	10	1	9	5	5	5	5
7	6	9	8	5	7	7	6	5	7	4	4	6	7	6	7
8	8	9	8	6	7	7	6	8	7	7	10	8	7	8	6

(continued)

(continued)

G1	G2	G3	G4	G16	G17	G18	G19	G24	G25	G26	G27	O1	O2	O3	O4
10	10	10	10	8	10	10	7	8	8	7	8	9	9	9	7
9	9	6	6	4	5	5	5	2	3	5	4	2	3	3	3
3	3	3	8	0	7	0	0	0	7	0	10	0	0	8	8
5	4	4	7	4	4	4	2	0	4	2	8	3	3	3	3
8	10	10	10	7	8	7	10	10	9	8	10	10	9	9	8
5	8	8	8	7	8	8	6	7	7	7	10	7	8	6	6
7	4	5	9	5	8	7	5	5	8	0	7	6	6	6	6
5	6	5	8	10	9	0	8	8	8	8	5	8	8	5	4
7	5	7	6	3	6	6	3	5	6	6	5	5	5	5	5
10	8	9	8	8	8	8	6	8	8	6	6	8	7	5	8
10	10	10	10	10	8	10	9	10	10	10	10	10	9	9	8
6	6	4	5	0	5	5	5	5	8	5	9	6	4	5	5
10	3	7	9	0	5	7	7	10	8	10	10	5	5	5	5
5	7	8	7	8	7	8	7	8	6	7	6	8	6	7	6
9	10	9	9	10	6	6	7	9	8	8	8	10	7	7	10
10	10	10	10	9	10	10	9	10	10	10	10	9	9	9	9
10	10	10	10	10	10	10	8	10	10	6	10	7	6	8	8
7	7	7	8	7	8	8	6	8	8	7	7	8	7	8	8
9	5	7	9	6	8	8	4	6	7	4	5	6	5	5	4
9	9	10	8	8	9	8	7	8	8	7	9	8	7	5	7
6	6	5	4	4	4	4	3	4	3	4	5	4	5	5	5
7	8	8	9	7	5	4	7	10	8	8	8	6	4	4	7
8	6	6	5	7	6	0	8	7	9	7	7	7	7	6	7
6	8	8	9	9	9	9	5	9	8	7	9	9	5	5	9
9	5	6	7	10	10	8	7	8	9	10	10	8	8	7	8
8	7	8	5	8	7	4	5	8	5	5	9	3	5	3	5
7	8	7	4	8	8	8	7	7	6	6	7	8	7	7	7
8	7	10	10	10	10	10	10	10	10	10	10	9	6	4	8
5	9	10	5	9	9	6	8	10	10	10	9	8	7	9	9
9	6	6	7	10	10	6	6	9	10	10	9	10	10	10	9
0	4	7	5	10	8	9	9	9	7	8	7	8	9	9	9
4	8	8	6	9	9	7	2	9	9	9	9	8	8	8	7
8	8	10	8	7	7	5	5	5	10	8	3	7	7	6	7
9	10	10	7	5	4	0	7	10	10	10	10	5	4	5	9
10	10	10	10	7	0	0	8	2	8	1	0	4	5	3	3
10	8	8	8	5	5	8	8	10	10	10	10	6	6	5	5
7	10	10	8	10	10	8	8	10	9	10	10	7	8	10	6
10	9	9	6	9	9	0	9	10	8	9	9	8	7	10	7
8	10	8	5	7	6	5	7	10	10	10	10	6	6	7	7
10	8	8	7	8	8	7	5	10	8	8	10	8	8	7	8
8	7	8	8	10	10	2	1	8	10	8	8	9	7	9	10
8	8	8	8	6	7	7	4	8	8	7	7	7	5	6	7
7	9	8	8	9	8	8	7	9	9	9	7	10	9	7	7
8	8	9	9	7	7	8	7	7	8	7	7	7	7	8	8
8	7	8	7	8	8	8	7	8	8	7	8	8	8	8	7
8	7	7	8	7	7	8	7	8	8	7	8	7	7	8	7

(continued)

(continued)

G1	G2	G3	G4	G16	G17	G18	G19	G24	G25	G26	G27	O1	O2	O3	O4
8	8	8	8	7	6	8	6	9	8	7	9	8	8	6	6
8	8	8	9	9	6	8	9	8	9	10	10	8	8	8	5
7	8	8	6	8	9	9	6	8	8	8	8	8	8	6	8
7	9	9	8	6	8	8	5	8	7	5	9	7	5	7	4
10	10	10	8	9	8	8	8	10	10	10	10	10	10	9	9
6	8	7	8	9	8	10	8	8	9	9	8	8	7	7	5
8	8	8	8	8	8	8	8	8	8	5	10	8	8	8	7
10	0	0	10	0	7	5	0	0	3	0	10	0	0	0	0
8	5	9	4	6	8	8	5	6	6	4	5	6	5	5	4
9	9	9	9	8	8	8	7	7	3	0	9	7	7	8	8
8	9	8	8	8	8	8	8	8	9	9	8	8	8	9	5
7	7	7	7	7	7	7	5	7	7	7	5	8	7	5	6
9	9	9	9	7	7	8	8	8	7	8	6	8	6	6	7
5	7	4	10	0	10	10	0	5	5	0	10	0	0	0	0
9	9	9	9	9	10	10	9	10	10	10	10	10	10	5	5
8	8	9	7	7	8	8	7	8	7	7	8	8	8	6	8
9	10	10	7	9	9	8	4	9	9	9	8	8	7	9	9
10	10	10	10	10	10	9	7	10	10	10	9	7	7	5	9
8	6	9	9	7	9	8	5	6	6	5	5	6	7	5	4
7	7	8	5	8	8	7	6	5	5	7	4	5	6	6	6
9	10	10	10	9	8	9	8	8	8	8	9	9	8	6	7
7	7	6	6	4	6	6	4	4	6	3	5	4	4	4	4
8	8	8	8	9	8	7	9	10	3	7	10	9	8	7	7
8	8	8	8	7	8	5	8	10	10	7	10	8	7	7	7
10	10	10	10	10	10	10	10	10	10	10	9	10	10	10	10
10	10	10	10	9	10	10	9	10	10	10	10	9	9	9	8
9	10	10	10	8	10	10	8	10	10	10	10	9	8	8	7
4	6	8	8	7	7	7	5	4	7	5	9	6	6	7	5
8	8	8	7	7	8	9	7	7	5	7	4	8	9	9	9
8	8	8	8	6	7	7	4	6	10	6	6	7	7	7	5
8	8	4	8	5	5	5	1	0	5	0	10	2	2	2	2
7	7	7	7	7	8	8	4	7	7	6	6	6	6	6	6
8	7	7	8	10	9	8	9	10	9	8	9	9	8	7	8
9	9	7	8	9	8	8	8	8	8	9	8	9	7	8	6
5	3	4	3	4	5	3	5	2	3	5	4	4	2	4	7
6	8	8	9	9	9	8	7	9	8	9	10	8	8	7	7
9	10	10	10	6	8	9	8	0	10	10	10	10	9	6	9
4	5	5	7	4	4	5	4	2	4	2	7	5	5	3	3
8	8	8	8	10	10	10	10	10	10	10	8	10	7	7	7
9	9	9	9	10	8	8	8	8	8	7	8	9	9	8	8
10	10	10	10	8	8	8	8	8	8	8	9	9	9	8	8
10	10	10	9	10	10	10	10	10	10	10	10	10	10	10	10
10	10	10	10	7	5	5	5	6	8	8	5	8	5	5	10
7	8	8	8	4	5	5	4	5	4	5	8	7	6	8	4
8	8	8	8	5	8	8	5	5	5	5	7	6	6	5	5
8	6	8	5	5	5	5	3	3	9	3	2	5	3	5	3

(continued)

(continued)

G1	G2	G3	G4	G16	G17	G18	G19	G24	G25	G26	G27	O1	O2	O3	O4
10	10	10	10	10	10	10	10	9	10	10	10	10	9	10	10
7	7	7	7	7	8	8	5	6	7	7	9	6	7	5	5
8	7	7	8	8	9	5	5	6	7	6	5	7	7	6	6
10	10	10	10	9	10	10	10	9	10	10	10	10	9	10	5
7	9	9	9	8	9	8	8	9	8	8	7	9	10	8	8
9	8	8	8	9	9	8	7	10	8	9	10	9	8	7	8
8	6	6	7	5	7	5	4	5	2	5	5	6	5	5	4
8	9	9	9	6	8	7	6	6	5	5	7	7	6	7	6
7	8	9	9	9	10	10	7	10	5	8	8	10	10	5	9
9	8	8	8	8	9	7	8	0	5	7	10	8	8	9	2
10	10	10	10	6	10	7	8	10	9	2	8	9	9	7	6
10	10	9	10	10	10	10	9	10	10	10	10	10	10	9	10
8	9	9	8	8	8	8	8	8	8	8	8	8	8	9	8
8	10	10	10	8	8	8	8	9	9	9	8	9	8	9	8
8	8	8	5	5	8	8	8	6	8	10	5	7	7	5	7
6	6	7	7	6	7	5	2	5	5	5	0	6	10	6	6
10	10	10	10	5	10	10	10	10	10	10	10	10	10	5	10
8	7	8	8	7	9	9	7	6	8	8	8	7	7	5	6
8	7	8	7	8	8	8	8	9	9	8	9	8	7	7	6
7	7	7	8	8	9	8	7	8	8	8	9	7	7	7	7
10	10	10	10	10	10	10	10	10	10	10	10	10	10	10	10
10	10	10	9	7	9	9	7	8	8	8	7	8	8	8	8
10	10	10	10	10	10	10	5	10	10	10	10	9	10	9	9
10	10	10	10	10	10	10	10	10	10	10	10	10	10	9	9
10	10	10	9	10	10	9	9	10	6	10	10	10	10	7	9
7	9	9	8	9	10	9	8	8	8	8	8	8	7	5	7
9	9	9	9	9	9	8	8	9	9	8	7	9	8	8	8
6	5	5	7	1	5	6	5	5	10	5	10	3	0	5	5
10	10	10	10	7	10	10	10	10	10	10	10	10	10	5	10
8	9	10	9	9	10	9	9	9	10	10	9	10	9	10	9
6	8	8	9	3	8	5	5	5	5	7	6	5	5	6	6
9	9	9	9	5	8	5	6	9	9	8	10	8	8	8	8
8	9	9	8	5	8	8	8	8	8	7	9	7	7	5	7
6	7	7	7	6	6	6	3	3	6	0	6	5	5	5	5
8	8	8	9	7	8	8	8	5	8	7	10	7	7	7	6
8	8	9	6	6	7	5	5	10	5	0	10	7	7	5	5
8	9	9	7	6	7	7	6	9	7	7	7	7	6	7	7
8	4	6	7	3	6	6	6	0	6	0	9	6	5	4	6
9	9	9	9	9	8	8	8	7	8	8	8	8	8	8	8
6	7	7	6	6	6	4	4	5	6	8	5	2	3	3	4
6	7	7	7	4	6	4	4	4	8	4	5	6	7	7	5
8	7	7	9	7	10	5	6	8	8	6	9	6	7	6	7
10	10	10	9	8	7	8	7	8	9	9	8	5	5	5	4
8	7	8	10	8	9	6	7	8	7	8	8	8	8	7	8
8	9	7	8	9	8	8	7	8	7	5	9	6	8	8	8

(continued)

(continued)

G1	G2	G3	G4	G16	G17	G18	G19	G24	G25	G26	G27	O1	O2	O3	O4
7	7	5	7	8	8	6	6	9	7	8	8	7	7	6	7
9	9	10	8	8	8	6	5	10	10	10	10	7	7	5	6
8	6	9	9	8	9	8	9	8	7	7	8	9	9	7	8
7	7	8	9	7	7	7	8	7	8	9	7	6	8	7	7
7	7	8	7	8	7	8	7	8	8	6	5	7	8	7	7
9	10	9	9	8	7	9	9	6	6	6	6	7	9	8	8
7	7	7	6	6	6	9	9	8	3	5	8	6	9	9	8
9	10	7	8	7	5	10	10	10	10	7	10	6	8	9	7
4	6	5	7	4	4	4	3	10	9	10	9	4	4	3	4
8	8	8	8	8	8	7	6	10	8	10	10	10	8	8	7
8	6	3	8	6	8	5	1	10	7	10	10	7	7	8	6
7	7	8	7	8	8	8	6	4	6	5	9	7	7	5	6
9	10	9	9	8	7	7	9	7	4	7	4	5	9	8	6
10	10	8	9	7	6	5	8	6	7	6	6	7	9	9	5
4	6	9	8	9	9	7	7	0	1	0	0	8	7	9	4
8	8	7	7	4	6	5	5	4	4	4	5	3	5	4	3
8	8	8	8	7	7	8	6	8	7	8	9	6	7	7	9
8	8	3	4	9	8	7	6	6	8	6	8	7	8	8	5
6	8	8	7	5	7	7	7	8	7	8	8	7	7	7	7
9	10	9	8	7	8	8	8	6	7	6	8	8	8	8	8
4	5	5	7	6	9	6	8	9	9	8	8	8	10	10	7
6	8	8	8	8	8	8	6	5	4	6	6	9	9	9	8
9	9	7	7	4	6	5	3	3	5	3	6	4	3	7	5
8	9	5	5	7	8	7	6	6	7	6	6	7	6	7	4
10	10	8	7	7	8	6	5	6	8	5	7	5	6	8	8
7	9	8	7	7	8	10	8	9	8	7	8	9	8	8	7
9	9	9	9	6	3	4	4	7	8	6	8	6	3	4	4
6	5	5	7	6	7	6	4	9	8	8	9	7	7	7	6
9	9	9	9	9	10	10	8	7	6	8	8	8	8	8	7
8	9	9	7	9	9	8	8	3	5	3	6	7	6	7	8
8	8	8	8	4	5	5	8	8	2	8	7	8	7	8	6
8	9	9	4	8	9	9	7	6	7	8	6	9	7	8	7
10	9	7	7	7	8	8	8	8	7	5	7	7	7	7	6

G gene, *O* outcome

References

1. Tsao DA, Chang HJ, Hsiung SK, Huang SE, Chang MS, Chiu HH, Chen YF, Cheng TL, Shiu-Ru L (2010) Gene expression profiles for predicting the efficacy of the anticancer drug 5-fluorouracil in breast cancer. DNA Cell Biol 29:285–293
2. Latan MS, Laddha NC, Latani J, Imran MJ, Begum R, Misra A (2012) Suppression of Cytokine gene expression and improved therapeutic efficacy of microemulsion- based tacrolimus cream for atopic dermatitis. Drug Deliv Transl Res 2:129–141
3. Albertin PS (1999) Longitudinal data analysis (repeated measures) in clinical trials. Stat Med 18:2863–2870

4. Yang X, Shen Q, Xu H, Shoptaw S (2007) Functional regression analysis using an F test for longitudinal data with large numbers of repeated measures. Stat Med 26:1552–1566
5. Sverdlov L (2001) The fastclus procedure as an effective way to analyze clinical data. In: SUGI proceedings 26, paper 224, Long Beach, CA
6. Gifi A (1990) Non linear multivariate analysis. Department of Data Theory, Leiden
7. Alpaydin E (2004) Introduction to machine learning. http://books.google.com. Accessed 25 June 2012
8. Van der Kooij AJ (2007) Prediction accuracy and stability of regression with optimal scaling transformations. Ph.D. thesis, Leiden University, Netherlands
9. Hojsgaard S, Halekoh U (2005) Overdispersion. Danish Institute of Agricultural Sciences. Copenhagen. http://gbi.agrsci.dk/statistics/courses. Accessed 18 Dec 2012
10. Wang L, Gordon MD, Zhu J (2006) Regularized least absolute deviations regression and an efficient algorithm for parameter tuning. In: Sixth international conference data mining 2006. doi:10.1109/ICDM.2006.134
11. Waaijenberg S, Zwinderman AH (2007) Penalized canonical correlation analysis to quantify the association between gene expression and DNA markers. BMC Proc 1(Suppl 1): S122–S125
12. Yoshiwara K, Tajima A, Yahata T, Kodama S, Fujiwara H et al (2010) Gene expression profile for predicting survival in advanced stage serous ovarian cancer across two independent data sets. PLoS One. doi:10.1371/journal.pone.0009615
13. Gururajan R, Quaddus M, Xu J (2008) Clinical usefulness of handheld wireless technology in healthcare. J Syst Info Technol 10:72–85
14. Kitsiou S, Manthou V, Vlachopoulou M, Markos A (2010) Adoption and sophistication of clinical information systems in geek public hospitals. 12th Med Confer Medical Biological Engineering 29:1011–1016
15. Hartmann A, Van der Kooij AJ, Zeeck A (2009) Models of clinical decision making by regression with optimal scaling. Psychother Res 19:482–492
16. Triantafilidou K, Venetis G, Markos A (2012) Short term results of autologous blood Injection for treatment of habitual TMJ luxation. J Craniofac Surg 23(3):689–692
17. Li Y (2008) Statistical methods in surrogate marker research. http>//deepblue.lib.umich.edu/handle. Accessed 18 Dec 2012
18. SPSS statistical software (2012) www.spss.com. Accessed 12 June 2012
19. Cleophas TJ, Zwinderman AH (2011) Statistics applied to clinical studies. Springer, New York
20. Barthelemew DJ (1995) Spearman and the origin and development of factor analysis. Br J Math Stat Psychol 48:211–220
21. Wold H (1966) Estimation of principle components and related models by iterative least squares. In: Krishnaiah PR (ed) Multivariate analysis. Academic Press, New York, pp 391–420
22. Sun L, Ji S, Yu S, Ye J (2009) On the equivalence between canonical correlation analysis and orthonormalized partial least squares. In: Proceeding of IJCAI conference on artificial intelligence 2009. Morgan Kaufman Publishers Jme, San Francisco, pp 1230–1235
23. Sherlock C, Roberts G (2009) Optimal scaling of random walk. Bernoulli 15:774–798

Chapter 4
Optimal Scaling: Regularization Including Ridge, Lasso, and Elastic Net Regression

1 Summary

1.1 Background

In the previous chapter we showed that linear regression, although commonly used in clinical research for analyzing the effects of predictors (x-variables) on outcome variables (y-axis variables), has a problem because consecutive levels of the predictor variables are assumed to be equal while in practice this is virtually never true. We showed that optimal scaling can largely improve the sensitivity of testing, but in order to fully benefit from this methodology the risk of overfitting, otherwise called overdispersion, of the data should be accounted.

1.2 Objective

In the current chapter we will address the subject of regularization, a method for correcting discretized variables for overdispersion.

1.3 Methods

Ridge regression, lasso regression, and elastic net regression will be demonstrated using the example from the previous chapter once more.

1.4 Results

The ridge optimal scaling model produced eight p-values < 0.01, while traditional regression and unregularized optimal scaling produced only 3 and 2 p-values < 0.01.

T.J. Cleophas and A.H. Zwinderman, *Machine Learning in Medicine*,
DOI 10.1007/978-94-007-5824-7_4, © Springer Science+Business Media Dordrecht 2013

Lasso optimal scaling eliminated 4 of 12 predictors from the analysis, while, of the remainder, only two were significant at $p < 0.01$. Similarly elastic net optimal scaling did not provide additional benefit.

1.5 Conclusions

1. Optimal scaling shows similarly sized effects compared to traditional regression. In order to benefit from optimal scaling a regularization procedure for the purpose of correcting overdispersion is desirable.
2. Ridge optimal scaling performed much better than did traditional regression giving rise to many more statistically significant predictors.
3. Lasso optimal scaling shrinks some b-values to zero, and is particularly suitable if you are looking for a limited number of strong predictors.
4. Elastic net optimal scaling works better than lasso if the number of predictors is larger than the number of observations.

2 Introduction

In the previous chapter we showed that linear regression, although commonly used in clinical research for analyzing the effects of predictors (x-variables) on outcome variables (y-axis variables), has a problem because consecutive levels of the predictor variables are assumed to be equal while in practice this is virtually never true. We showed that optimal scaling can largely improve the sensitivity of testing, but in order to fully benefit from this methodology the risk of overdispersion of the data should be accounted. Solutions were given by Robert Tibshirani, professor of statistics at Stanford University, Palo Alto CA, 1996, who published about the possibility to shrink the regression coefficients in order to further improve the fit of the regression model as used [1].

In the current chapter, using the example from the previous chapter, we will address the subject of regularization, a method for correcting discretized variables for overfitting, otherwise called overdispersion.

3 Some Terminology

3.1 Discretization

Converting continuous variables into discretized values in a regression model.

3.2   Splines

Cut pieces of a non-linear graph, originally thin wooden strips for modeling cars and airplanes.

3.3   Overdispersion, Otherwise Called Overfitting

The phenomenon that the spread in the data is wider than compatible with Gaussian modeling. This phenomenon is particularly common with discretization of continuous variables.

3.4   Regularization

Correcting discretized variables for overfitting, otherwise called overdispersion.

3.5   Ridge Regression

Important method for shrinking b-values for the purpose of adjusting overdispersion.

3.6   Monte Carlo Methods

Iterative testing in order to find the best fit solution for a statistical problem.

3.7   Cross-Validation

Splitting the data into a k-fold scale and comparing it with a k-1 fold scale.

3.8   Lasso Regression

Shrinking procedure slightly different from ridge regression, because it shrinks the smallest b-values to 0.

3.9 *Elastic Net Regression*

Shrinking procedure similar to lasso, but made suitable for larger numbers of predictors.

4   Some Theory

Optimal scaling makes use of processes like discretization (converting continuous variables into discretized values), and regularization (correcting discretized variables for overfitting, otherwise called overdispersion) [2–5]. Discretization was addressed in the previous chapter.

Regularization has to take place for the purpose of correcting overdispersed models. Generally, the standard error is increased. Various methods for adjustment are possible. Hojsgaard and Halekoh [4] recommended to use the [chi-square/degrees of freedom] ratio for adjustment. But shrinking the b-value is also adequate. Ridge regression is an important approach. The regression coefficient b is minimized by a shrinking factor λ such that $b_{ridge} = b/(1 + \lambda)$, and that, with $\lambda = 0$, $b_{ridge} = b$, and, with $\lambda = \infty$, $b_{ridge} = 0$. Calculations are based on likelihood statistics adjusted for degrees of freedom, and it seems true that there always exists a value for λ such that it provides a better scale model than did the traditional linear model. Knowing this, one elegant approach is the Monte Carlo approach, i.e. perform multiple tests in order to find the best fit scale. For the purpose cross-validation splitting the data into a k-fold scale and comparing it with a k-1 fold scale is a common method. In addition to ridge regression, SPSS [6] offers lasso regression and elastic net regression. Lasso although it uses similarly sized factors to reduce the size of b-values, it shrinks the smallest b-values in your data to 0, thereby eliminating some variables. This improves the prediction accuracy, particularly if you are looking for a model with a limited number of strong predictors. In contrast, if you are looking for a complex model with a large number of predictors albeit weak ones, then ridge regression will perform better. Elastic net regression is like lasso but performs better if the number of predictors is larger than the number of observations.

5   Example

The 250 patients' data-file from the previous chapter was used once more (Appendix). It was supposed to include 27 variables consistent of both patients' microarray gene expression levels and their drug efficacy scores. The data file is in the appendix. All variables were standardized by scoring them on 11 points linear scales (0–10). The following genes were highly

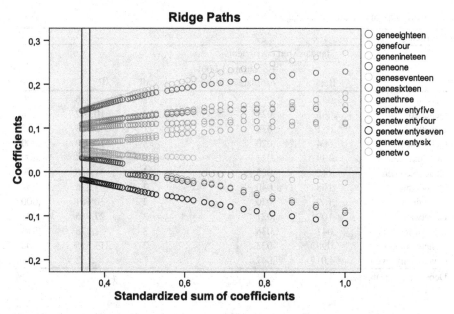

Fig. 4.1 Optimal scaling met ridge regression. The graph shows the adjusted b-values of the best fit scale model (*left vertical line*), the b-values are also in the Table 4.1. The graph shows how the b-value of different predictors gradually increase as the shrinking factor λ decreases (*from the left to right end of the graph*). The right vertical line is the situation where the spread in the data has increased by one standard error above the best model (*left line*), and this model has thus deteriorated correspondingly

expressed: the genes 1–4, 16–19, and 24–27. As outcome variable composite scores of the variables 20–23 were used.

The optimally scaled model without regularization shows similarly sized effects compared to the traditional regression model although the overall r-square was slightly larger with a value of 0.736. In order to fully benefit from optimal scaling a regularization procedure for the purpose of correcting overdispersion is needed. A ridge path model will be first used. SPSS statistical software is used [6].

Command: Analyze….Regression….Optimal Scaling….Dependent Variable: Var 28 (Define Scale: mark spline ordinal 2.2)….Independent Variables: Var 1, 2, 3, 4, 16, 17, 18, 19, 24, 25, 26, 27 (all of them Define Scale: mark spline ordinal 2.2)….Discretize: Method Grouping, Number categories 7)….click Regularization….mark Ridge….OK.

Figure 4.1 shows the adjusted b-values of the best fit Ridge scale model (left vertical line), the b-values are also in Table 4.1. Figure 4.1 shows how the b-values of the different predictors gradually increase as the factor λ decreases. The right vertical line is the situation where the spread in the data has increased by one standard error above the best model (left line), and where the model has, thus,

Table 4.1 Optimal scaling with ridge regression

Coefficients

	Standardized coefficients		df	F	Sig.
	Beta	Bootstrap (1,000) estimate of Std. error			
Geneone	,032	,033	2	,946	,390
Genetwo	,068	,021	3	10,842	,000
Genethree	,051	,030	1	2,963	,087
Genefour	,064	,020	3	10,098	,000
Genesixteen	,139	,024	4	34,114	,000
Geneseventeen	,142	,025	2	31,468	,000
Geneeighteen	,108	,040	2	7,236	,001
Genenineteen	,109	,020	2	30,181	,000
Genetwentyfour	,109	,021	2	27,855	,000
Genetwentyfive	,041	,038	3	1,178	,319
Genetwentysix	,098	,023	2	17,515	,000
Genetwentyseven	−,017	,047	1	,132	,716

Dependent variable: 20–23

deteriorated correspondingly. The sensitivity of this model is much better than the traditional regression with 8 p-values < 0.01, while the traditional and unregularized Optimal Scaling only produced 3 and 2 p-values < 0.01.

Also the lasso regularization model is possible (Var = variable).

Command: Analyze….Regression….Optimal Scaling….Dependent Variable: Var 28 (Define Scale: mark spline ordinal 2.2)….Independent Variables: Var 1, 2, 3, 4, 16, 17, 18, 19, 24, 25, 26, 27 (all of them Define Scale: mark spline ordinal 2.2)…. Discretize: Method Grouping, Number categories 7)….click Regularization…. mark Lasso….OK.

Figure 4.2 shows the adjusted b-values of the best fit lasso scale model (left vertical line), the b-values are also in Table 4.2. The b-values of the genes 1, 3, 25 and 27 are now shrunk to zero, and eliminated from the analysis. Lasso is particularly suitable if you are looking for a limited number of predictors and improves prediction accuracy by leaving out weak predictors.

Finally, the elastic net method is applied.

Command: Analyze….Regression….Optimal Scaling….Dependent Variable: Var 28 (Define Scale: mark spline ordinal 2.2)….Independent Variables: Var 1, 2, 3, 4, 16, 17, 18, 19, 24, 25, 26, 27 (all of them Define Scale: mark spline ordinal 2.2)….Discretize: Method Grouping, Number categories 7)….click Regularization….mark Elastic Net….OK.

Table 4.3 gives the results, that are, in our example, pretty much the same, as it is with lasso. Elastic net does not provide additional benefit in this example but works better than Lasso if the number of predictors is larger than the number of observations.

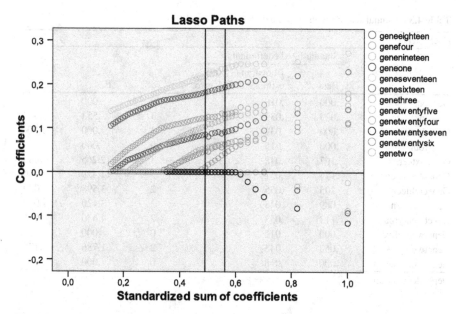

Fig. 4.2 Lasso regression. The graph shows the adjusted b-values of the best fit scale model (*left vertical line*), the b-values are also in the Table 4.2. The graph shows how the b-value of different predictors gradually increase as the shrinking factor λ decreases (*from the left to right end of the graph*). The right vertical line is the situation where the spread in the data has increased by one standard error above the best model (*left line*), and this model has thus deteriorated correspondingly

Table 4.2 Optimal scaling with lasso regression

Coefficients

| | Standardized coefficients | | | | |
	Beta	Bootstrap (1,000) estimate of Std. error	df	F	Sig.
Geneone	,000	,020	0	,000	.
Genetwo	,054	,046	3	1,390	,247
Genethree	,000	,026	0	,000	.
Genefour	,011	,036	3	,099	,960
Genesixteen	,182	,084	4	4,684	,001
Geneseventeen	,219	,095	3	5,334	,001
Geneeighteen	,086	,079	2	1,159	,316
Genenineteen	,105	,063	2	2,803	,063
Genetwentyfour	,124	,078	2	2,532	,082
Genetwentyfive	,000	,023	0	,000	.
Genetwentysix	,048	,060	2	,647	,525
Genetwentyseven	,000	,022	0	,000	.

Dependent variable: 20–23

Table 4.3 Optimal scaling with elastic net (ridge 0.00 t/m/1.00)

Coefficients					
	Standardized coefficients				
	Beta	Bootstrap (1,000) estimate of Std. error	df	F	Sig.
Geneone	,000	,016	0	,000	.
Genetwo	,029	,039	3	,553	,647
Genethree	,000	,032	3	,000	1,000
Genefour	,000	,015	0	,000	.
Genesixteen	,167	,048	4	12,265	,000
Geneseventeen	,174	,051	3	11,429	,000
Geneeighteen	,105	,055	2	3,598	,029
Genenineteen	,089	,048	3	3,420	,018
Genetwentyfour	,113	,053	2	4,630	,011
Genetwentyfive	,000	,012	0	,000	.
Genetwentysix	,062	,046	2	1,786	,170
Genetwentyseven	,000	,018	0	,000	.

Dependent variable: 20–23

6   Discussion

Traditional linear regression has been demonstrated not to perform well in case of multiple independent variables. A sharp increase of the t-values of some x-values is often observed if other x-values are removed. This phenomenon is called instable regression coefficients, sometimes referred to as "bouncing betas" [3], and arises when predictors are correlated or when a large number of predictors relative to the number of observations is present. Shrinking the regression coefficients has been demonstrated to be beneficial not only to counterbalance overdispersion but also to reduce this phenomenon of instability [4, 5].

Optimally scaled modeling showed similarly sized effects compared to the traditional linear regression. In order to fully benefit from optimal scaling a regularization procedure for the purpose of correcting overdispersion appeared to be desirable.

Particularly, the sensitivity of the ridge optimal scaling was better for the purpose than that of traditional linear regression giving rise to many more significant predictors in your model.

Lasso optimal scaling shrunk some variable b-values to zero, and is, therefore, particularly suitable if you are looking for a limited number of strong predictors. Elastic net optimal scaling worked better than lasso, and may be particularly suitable if the number of predictors is larger than the number of observations.

We hope this paper will stimulate clinical investigators to start using this optimized analysis method for predictive trials.

7 Conclusions

1. Optimally scaled modeling shows similarly sized effects compared to the traditional linear regression. In order to fully benefit from optimal scaling a regularization procedure for the purpose of correcting overdispersion is desirable.
2. Particularly, the sensitivity of the ridge optimal scaling may be better than that of traditional linear regression giving rise to more significant predictors in the data.
3. Lasso optimal scaling shrinks some variable b-values to zero, and is, therefore, particularly suitable if you are looking for a limited number of strong predictors.
4. Elastic net optimal scaling works better than lasso if the number of predictors is larger than the number of observations.

8 Appendix: Datafile of 250 Subjects Used as Example

G1	G2	G3	G4	G16	G17	G18	G19	G24	G25	G26	G27	O1	O2	O3	O4
8	8	9	5	7	10	5	6	9	9	6	6	6	7	6	7
9	9	10	9	8	8	7	8	8	9	8	8	8	7	8	7
9	8	8	8	8	9	7	8	9	8	9	9	9	8	8	8
8	9	8	9	6	7	6	4	6	6	5	5	7	7	7	6
10	10	8	10	9	10	10	8	8	9	9	9	8	8	8	7
7	8	8	8	8	7	6	5	7	8	8	7	7	6	6	7
5	5	5	5	5	6	4	5	5	6	6	5	6	5	6	4
9	9	9	9	8	8	8	8	9	8	3	8	8	8	8	8
9	8	9	8	9	8	7	7	7	7	5	8	8	7	6	6
10	10	10	10	10	10	10	10	10	8	8	10	10	10	9	10
2	2	8	5	7	8	8	8	9	3	9	8	7	7	7	6
7	8	8	7	8	6	6	7	8	8	8	7	8	7	8	8
8	9	9	8	10	8	8	7	8	8	9	9	7	7	8	8
7	7	8	8	8	9	10	7	9	4	8	8	9	8	7	7
3	4	3	8	4	4	4	3	4	3	4	4	4	4	3	4
7	8	8	5	8	8	7	6	7	7	8	7	10	8	8	7
8	8	8	8	6	8	5	1	9	7	7	8	7	7	8	6
7	8	8	8	8	9	8	7	10	10	9	8	9	9	9	9
8	4	3	8	3	5	5	3	2	10	1	0	5	3	4	3
8	7	6	10	8	8	7	6	4	4	5	5	7	7	7	5
9	9	10	8	8	9	7	7	8	9	8	9	8	7	8	7
6	6	6	6	4	5	4	5	3	9	3	4	4	5	4	3
8	8	8	7	7	7	8	6	8	7	9	4	6	7	8	9
9	9	10	9	10	10	7	10	10	10	10	10	8	8	8	5
8	7	8	8	9	8	9	8	8	8	8	8	8	8	8	9
8	5	5	4	2	1	1	0	0	1	0	0	3	2	4	5
6	6	6	6	5	6	3	5	4	4	4	5	5	6	3	4
7	8	9	8	8	9	9	6	9	8	8	10	9	8	7	7
8	8	8	7	7	7	7	6	7	8	7	8	7	6	6	6
8	8	8	8	9	8	9	8	9	8	9	9	9	8	7	8

(continued)

(continued)

G1	G2	G3	G4	G16	G17	G18	G19	G24	G25	G26	G27	O1	O2	O3	O4
7	7	7	6	7	7	9	7	7	7	7	8	8	6	7	7
9	9	9	9	6	9	8	7	8	8	8	9	8	8	8	8
10	10	10	10	9	9	10	5	10	2	9	9	8	10	8	8
9	8	9	9	8	7	7	8	9	9	9	9	8	5	9	7
8	9	9	9	8	7	7	6	7	8	8	8	8	7	8	6
3	4	2	5	4	2	2	4	4	4	3	4	6	2	3	2
8	8	9	9	8	8	8	8	8	8	8	8	8	8	8	8
8	6	7	6	7	7	8	6	7	6	5	5	6	7	7	6
10	10	10	10	7	10	10	8	10	10	10	10	10	8	8	7
8	10	9	8	8	8	7	6	7	7	10	8	9	8	8	7
8	8	8	8	8	8	7	8	7	8	8	8	9	9	8	7
5	7	7	8	5	7	7	3	1	6	3	10	5	6	6	5
10	9	9	10	7	9	9	9	9	9	9	8	8	9	7	7
9	7	7	9	3	6	4	2	1	8	2	1	6	6	6	6
8	8	10	8	9	8	7	8	8	7	8	8	9	6	5	7
6	8	8	8	9	10	10	9	10	9	9	10	9	8	5	5
8	8	8	8	10	8	7	10	8	8	7	10	9	7	8	6
6	5	5	6	6	6	4	6	3	5	0	3	7	5	5	3
9	9	9	8	8	9	8	7	6	7	8	10	8	8	8	6
9	10	8	8	9	10	10	9	7	8	9	7	8	8	7	7
8	8	8	9	6	8	7	6	8	9	8	8	7	7	6	5
8	5	6	7	8	8	7	7	4	6	7	6	8	8	7	6
4	1	4	9	0	0	7	0	0	10	0	10	0	0	0	0
5	5	7	5	7	7	8	5	7	7	5	5	7	7	7	7
5	5	6	5	4	4	4	3	3	2	3	3	3	4	3	3
7	9	9	10	5	9	9	9	9	6	7	6	10	7	10	9
10	10	10	10	8	9	9	6	7	8	8	10	7	7	7	6
8	8	8	8	6	9	8	7	7	6	6	2	7	7	7	5
6	6	7	9	8	8	7	6	1	9	0	4	6	7	7	6
6	7	7	7	6	5	5	5	5	7	3	5	7	6	6	8
9	9	9	9	8	8	9	6	8	7	6	10	8	7	7	8
7	7	7	7	6	8	8	6	7	7	7	8	6	6	5	10
9	7	8	9	8	10	8	9	8	9	7	9	7	7	8	3
8	9	9	8	7	8	7	8	8	6	7	8	7	8	7	6
8	8	8	8	6	8	8	5	8	9	8	7	7	7	6	5
7	7	7	7	4	5	6	6	3	6	7	7	1	5	6	5
9	10	9	9	8	9	8	8	9	8	9	9	8	7	8	8
8	9	9	8	8	8	8	7	8	7	8	9	6	6	5	6
7	8	8	8	6	7	7	6	8	5	7	7	7	6	7	5
4	2	2	6	5	5	4	4	6	4	3	2	4	6	7	2
5	5	7	5	5	5	5	2	2	9	5	5	4	5	5	4
9	9	10	9	7	8	7	8	8	9	8	8	8	8	6	9
8	8	8	8	7	7	7	9	8	9	7	8	7	7	5	6
8	8	9	8	8	9	5	9	8	5	7	6	8	8	8	6
9	9	9	9	6	8	8	4	7	5	6	6	7	7	8	8

(continued)

(continued)

G1	G2	G3	G4	G16	G17	G18	G19	G24	G25	G26	G27	O1	O2	O3	O4
9	8	8	8	7	9	9	9	10	10	10	10	10	9	7	10
9	9	9	8	8	8	8	7	7	7	7	7	8	8	8	7
8	5	7	9	2	8	8	2	9	10	1	9	5	5	5	5
7	6	9	8	5	7	7	6	5	7	4	4	6	7	6	7
8	8	9	8	6	7	7	6	8	7	7	10	8	7	8	6
10	10	10	10	8	10	10	7	8	8	7	8	9	9	9	7
9	9	6	6	4	5	5	5	2	3	5	4	2	3	3	3
3	3	3	8	0	7	0	0	0	7	0	10	0	0	8	8
5	4	4	7	4	4	4	2	0	4	2	8	3	3	3	3
8	10	10	10	7	8	7	10	10	9	8	10	10	9	9	8
5	8	8	8	7	8	8	6	7	7	7	10	7	8	6	6
7	4	5	9	5	8	7	5	5	8	0	7	6	6	6	6
5	6	5	8	10	9	0	8	8	8	8	5	8	8	5	4
7	5	7	6	3	6	6	3	5	6	6	5	5	5	5	5
10	8	9	8	8	8	8	6	8	8	6	6	8	7	5	8
10	10	10	10	10	10	8	10	9	10	10	10	10	9	9	8
6	6	4	5	0	5	5	5	5	8	5	9	6	4	5	5
10	3	7	9	0	5	7	7	10	8	10	10	5	5	5	5
5	7	8	7	8	7	8	7	8	6	7	6	8	6	7	6
9	10	9	9	10	6	6	7	9	8	8	8	10	7	7	10
10	10	10	10	9	10	10	9	10	10	10	10	9	9	9	9
10	10	10	10	10	10	10	8	10	10	6	10	7	6	8	8
7	7	7	8	7	8	8	6	8	8	7	7	8	7	8	8
9	5	7	9	6	8	8	4	6	7	4	5	6	5	5	4
9	9	10	8	8	9	8	7	8	8	7	9	8	7	5	7
6	6	5	4	4	4	4	3	4	3	4	5	4	5	5	5
7	8	8	9	7	5	4	7	10	8	8	8	6	4	4	7
8	6	6	5	7	6	0	8	7	9	7	7	7	7	6	7
6	8	8	9	9	9	9	5	9	8	7	9	9	5	5	9
9	5	6	7	10	10	8	7	8	9	10	10	8	8	7	8
8	7	8	5	8	7	4	5	8	5	5	9	3	5	3	5
7	8	7	4	8	8	8	7	7	6	6	7	8	7	7	7
8	7	10	10	10	10	10	10	10	10	10	10	9	6	4	8
5	9	10	5	9	9	6	8	10	10	10	9	8	7	9	9
9	6	6	7	10	10	6	6	9	10	10	9	10	10	10	9
0	4	7	5	10	8	9	9	9	7	8	7	8	9	9	9
4	8	8	6	9	9	7	2	9	9	9	9	8	8	8	7
8	8	10	8	7	7	5	5	5	10	8	3	7	7	6	7
9	10	10	7	5	4	0	7	10	10	10	10	5	4	5	9
10	10	10	10	7	0	0	8	2	8	1	0	4	5	3	3
10	8	8	8	5	5	8	8	10	10	10	10	6	6	5	5
7	10	10	8	10	10	8	8	10	9	10	10	7	8	10	6
10	9	9	6	9	9	0	9	10	8	9	9	8	7	10	7
8	10	8	5	7	6	5	7	10	10	10	10	6	6	7	7
10	8	8	7	8	8	7	5	10	8	8	10	8	8	7	8

(continued)

(continued)

G1	G2	G3	G4	G16	G17	G18	G19	G24	G25	G26	G27	O1	O2	O3	O4
8	7	8	8	10	10	2	1	8	10	8	8	9	7	9	10
8	8	8	8	6	7	7	4	8	8	7	7	7	5	6	7
7	9	8	8	9	8	8	7	9	9	9	7	10	9	7	7
8	8	9	9	7	7	8	7	7	8	7	7	7	7	8	8
8	7	8	7	8	8	8	7	8	8	7	8	8	8	8	7
8	7	7	8	7	7	8	7	8	8	7	8	7	7	8	7
8	8	8	8	7	6	8	6	9	8	7	9	8	8	6	6
8	8	8	9	9	6	8	9	8	9	10	10	8	8	8	5
7	8	8	6	8	9	9	6	8	8	8	8	8	8	6	8
7	9	9	8	6	8	8	5	8	7	5	9	7	5	7	4
10	10	10	8	9	8	8	8	10	10	10	10	10	10	9	9
6	8	7	8	9	8	10	8	8	9	9	8	8	7	7	5
8	8	8	8	8	8	8	8	8	8	5	10	8	8	8	7
10	0	0	10	0	7	5	0	0	3	0	10	0	0	0	0
8	5	9	4	6	8	8	5	6	6	4	5	6	5	5	4
9	9	9	9	8	8	8	7	7	3	0	9	7	7	8	8
8	9	8	8	8	8	8	8	8	9	9	8	8	8	9	5
7	7	7	7	7	7	7	5	7	7	7	5	8	7	5	6
9	9	9	9	7	7	8	8	8	7	8	6	8	6	6	7
5	7	4	10	0	10	10	0	5	5	0	10	0	0	0	0
9	9	9	9	9	10	10	9	10	10	10	10	10	10	5	5
8	8	9	7	7	8	8	7	8	7	7	8	8	8	6	8
9	10	10	7	9	9	8	4	9	9	9	8	8	7	9	9
10	10	10	10	10	10	9	7	10	10	10	9	7	7	5	9
8	6	9	9	7	9	8	5	6	6	5	5	6	7	5	4
7	7	8	5	8	8	7	6	5	5	7	4	5	6	6	6
9	10	10	10	9	8	9	8	8	8	8	9	9	8	6	7
7	7	6	6	4	6	6	4	4	6	3	5	4	4	4	4
8	8	8	8	9	8	7	9	10	3	7	10	9	8	7	7
8	8	8	8	7	8	5	8	10	10	7	10	8	7	7	7
10	10	10	10	10	10	10	10	10	10	10	9	10	10	10	10
10	10	10	10	9	10	10	9	10	10	10	10	9	9	9	8
9	10	10	10	8	10	10	8	10	10	10	10	9	8	8	7
4	6	8	8	7	7	7	5	4	7	5	9	6	6	7	5
8	8	8	7	7	8	9	7	7	5	7	4	8	9	9	9
8	8	8	8	6	7	7	4	6	10	6	6	7	7	7	5
8	8	4	8	5	5	5	1	0	5	0	10	2	2	2	2
7	7	7	7	7	8	8	4	7	7	6	6	6	6	6	6
8	7	7	8	10	9	8	9	10	9	8	9	9	8	7	8
9	9	7	8	9	8	8	8	8	8	9	8	9	7	8	6
5	3	4	3	4	5	3	5	2	3	5	4	4	2	4	7
6	8	8	8	9	9	8	7	9	8	9	10	8	8	7	7
9	10	10	10	6	8	9	8	0	10	10	10	10	9	6	9
4	5	5	7	4	4	5	4	2	4	2	7	5	5	3	3
8	8	8	8	10	10	10	10	10	10	10	8	10	7	7	7

(continued)

(continued)

G1	G2	G3	G4	G16	G17	G18	G19	G24	G25	G26	G27	O1	O2	O3	O4
9	9	9	9	10	8	8	8	8	8	7	8	9	9	8	8
10	10	10	10	8	8	8	8	8	8	8	9	9	9	8	8
10	10	10	9	10	10	10	10	10	10	10	10	10	10	10	10
10	10	10	10	7	5	5	5	6	8	8	5	8	5	5	10
7	8	8	8	4	5	5	4	5	4	5	8	7	6	8	4
8	8	8	8	5	8	8	5	5	5	5	7	6	6	5	5
8	6	8	5	5	5	5	3	3	9	3	2	5	3	5	3
10	10	10	10	10	10	10	10	9	10	10	10	10	9	10	10
7	7	7	7	7	8	8	5	6	7	7	9	6	7	5	5
8	7	7	8	8	9	5	5	6	7	6	5	7	7	6	6
10	10	10	10	9	10	10	10	9	10	10	10	10	9	10	5
7	9	9	9	8	9	8	8	9	8	8	7	9	10	8	8
9	8	8	8	9	9	8	7	10	8	9	10	9	8	7	8
8	6	6	7	5	7	5	4	5	2	5	5	6	5	5	4
8	9	9	9	6	8	7	6	6	5	5	7	7	6	7	6
7	8	9	9	9	10	10	7	10	5	8	8	10	10	5	9
9	8	8	8	8	9	7	8	0	5	7	10	8	8	9	2
10	10	10	10	6	10	7	8	10	9	2	8	9	9	7	6
10	10	9	10	10	10	10	9	10	10	10	10	10	10	9	10
8	9	9	8	8	8	8	8	8	8	8	8	8	8	9	8
8	10	10	10	8	8	8	8	9	9	9	8	9	8	9	8
8	8	8	5	5	8	8	8	6	8	10	5	7	7	5	7
6	6	7	7	6	7	5	2	5	5	5	0	6	10	6	6
10	10	10	10	5	10	10	10	10	10	10	10	10	10	5	10
8	7	8	8	7	9	9	7	6	8	8	8	7	7	5	6
8	7	8	7	8	8	8	8	9	9	8	9	8	7	7	6
7	7	7	8	8	9	8	7	8	8	8	9	7	7	7	7
10	10	10	10	10	10	10	10	10	10	10	10	10	10	10	10
10	10	10	9	7	9	9	7	8	8	8	7	8	8	8	8
10	10	10	10	10	10	10	5	10	10	10	10	9	10	9	9
10	10	10	10	10	10	10	10	10	10	10	10	10	10	9	9
10	10	10	9	10	10	9	9	10	6	10	10	10	10	7	9
7	9	9	8	9	10	9	8	8	8	8	8	8	7	5	7
9	9	9	9	9	9	8	8	9	9	8	7	9	8	8	8
6	5	5	7	1	5	6	5	5	10	5	10	3	0	5	5
10	10	10	10	7	10	10	10	10	10	10	10	10	10	5	10
8	9	10	9	9	10	9	9	9	10	10	9	10	9	10	9
6	8	8	9	3	8	5	5	5	5	7	6	5	5	6	6
9	9	9	9	5	8	5	6	9	9	8	10	8	8	8	8
8	9	9	8	5	8	8	8	8	8	7	9	7	7	5	7
6	7	7	7	6	6	6	3	3	6	0	6	5	5	5	5
8	8	8	9	7	8	8	8	5	8	7	10	7	7	7	6
8	8	9	6	6	7	5	5	10	5	0	10	7	7	5	5
8	9	9	7	6	7	7	6	9	7	7	7	7	6	7	7
8	4	6	7	3	6	6	6	0	6	0	9	6	5	4	6

(continued)

(continued)

G1	G2	G3	G4	G16	G17	G18	G19	G24	G25	G26	G27	O1	O2	O3	O4
9	9	9	9	9	8	8	8	7	8	8	8	8	8	8	8
6	7	7	6	6	6	4	4	5	6	8	5	2	3	3	4
6	7	7	7	4	6	4	4	4	8	4	5	6	7	7	5
8	7	7	9	7	10	5	6	8	8	6	9	6	7	6	7
10	10	10	9	8	7	8	7	8	9	9	8	5	5	5	4
8	7	8	10	8	9	6	7	8	7	8	8	8	8	7	8
8	9	7	8	9	8	8	7	8	7	5	9	6	8	8	8
7	7	5	7	8	8	6	6	9	7	8	8	7	7	6	7
9	9	10	8	8	8	6	5	10	10	10	10	7	7	5	6
8	6	9	9	8	9	8	9	8	7	7	8	9	9	7	8
7	7	8	9	7	7	7	8	7	8	9	7	6	8	7	7
7	7	8	7	8	7	8	7	8	8	6	5	7	8	7	7
9	10	9	9	8	7	9	9	6	6	6	6	7	9	8	8
7	7	7	6	6	6	9	9	8	3	5	8	6	9	9	8
9	10	7	8	7	5	10	10	10	10	7	10	6	8	9	7
4	6	5	7	4	4	4	3	10	9	10	9	4	4	3	4
8	8	8	8	8	8	7	6	10	8	10	10	10	8	8	7
8	6	3	8	6	8	5	1	10	7	10	10	7	7	8	6
7	7	8	7	8	8	8	6	4	6	5	9	7	7	5	6
9	10	9	9	8	7	7	9	7	4	7	4	5	9	8	6
10	10	8	9	7	6	5	8	6	7	6	6	7	9	9	5
4	6	9	8	9	9	7	7	0	1	0	0	8	7	9	4
8	8	7	7	4	6	5	5	4	4	4	5	3	5	4	3
8	8	8	8	7	7	8	6	8	7	8	9	6	7	7	9
8	8	3	4	9	8	7	6	6	8	6	8	7	8	8	5
6	8	8	7	5	7	7	7	8	7	8	8	7	7	7	7
9	10	9	8	7	8	8	8	6	7	6	8	8	8	8	8
4	5	5	7	6	9	6	8	9	9	8	8	8	10	10	7
6	8	8	8	8	8	8	6	5	4	6	6	9	9	9	8
9	9	7	7	4	6	5	3	3	5	3	6	4	3	7	5
8	9	5	5	7	8	7	6	6	7	6	6	7	6	7	4
10	10	8	7	7	8	6	5	6	8	5	7	5	6	8	8
7	9	8	7	7	8	10	8	9	8	7	8	9	8	8	7
9	9	9	9	6	3	4	4	7	8	6	8	6	3	4	4
6	5	5	7	6	7	6	4	9	8	8	9	7	7	7	6
9	9	9	9	9	10	10	8	7	6	8	8	8	8	8	7
8	9	9	7	9	9	8	8	3	5	3	6	7	6	7	8
8	8	8	8	4	5	5	8	8	2	8	7	8	7	8	6
8	9	9	4	8	9	9	7	6	7	8	6	9	7	8	7
10	9	7	7	7	8	8	8	8	7	5	7	7	7	7	6

G gene, O outcome

References

1. Tibshirani R (1996) Regression shrinkage and selection via the lasso. J R Stat Soc 58: 267–288
2. Alpaydin E (2004) Introduction to machine learning. http://books.google.com. Accessed 25 June 2012
3. Van der Kooij AJ (2007) Prediction accuracy and stability of regression with optimal scaling transformations. Ph.D thesis Leiden University, Netherlands
4. Hojsgaard S, Halekoh U (2005) Overdispersion. Danish Institute of Agricultural Sciences, Copenhagen. http://gbi.agrsci.dk/statistics/courses. Accessed 18 Dec 2012
5. Wang L, Gordon MD, Zhu J (2006) Regularized least absolute deviations regression and an efficient algorithm for parameter tuning. 6th Int Conf Data Min. doi:10.1109/ICDM.2006.134
6. SPSS statistical software (2012) www.spss.com. Accessed 12 June 2012

Chapter 5
Partial Correlations

1 Summary

1.1 Background

The outcome of clinical research is generally affected by many more factors than a single one, and multiple regression assumes that these factors act independently of one another, but why should they not affect one another.

1.2 Objective

To assess the performance of partial regression analysis for the assessment of clinical trials with interaction between predictor factors.

1.3 Methods

A simulated 64 patient study of the effects of exercise on weight loss with calorie intake as covariate and a significant interaction on the outcome between the covariates.

1.4 Results

The simple linear correlations of weight loss versus exercise and versus calorie intake were respectively 0.41 (p=0.001) and −0.30 (p=0.015). Multiple linear regression adjusted for interaction showed that exercise was no longer a significant predictor of weight loss (regression coefficient=−0.24, p=0.807). By controlling for calorie intake using partial correlation analysis, the correlation coefficient between exercise and weight loss increased from 0.41 (p=0.001) to 0.60 (p=0.0001), 48% increase. The interaction, obviously, caused a 48% underestimation of the correlation between exercise and weight loss.

1.5   Conclusions

1. Multiple regression is a biased methodology, if subgroup effects interact on the outcome with one another, because it can only adjust confounding but not interaction.
2. Partial correlation analysis can not demonstrate the amount of interaction in the data, but it can remove its effect, and, thus, establish what would have happened, had there been no interaction.

2   Introduction

Nowadays we have come a long way to recognize that clinical factors are affected by many more factors than a single one, and multiple regression assumes that all of these factors act independently of one another, but why should that be true [1]. If all of these factors affect the outcome, why should they not affect one another. The point is, that, if you don't have an adequate method to study this issue, you will never know.

As an example, both calorie intake and exercise are predictors of weight loss. However, exercise makes you hungry and patients on weight training may be inclined to reduce their calorie intake. So, there may be an interaction between calorie intake and exercise on weight loss. Suppose an interaction variable (x=calorie intake * exercise, with * symbol of multiplication) is included in the analysis, and a significant interaction is established. With a significant interaction the overall analysis of these data is rather meaningless. The best method to find the true effect of calorie intake on weight loss or of exercise on weight loss, would be to repeat the study with one of the two variables held constant and the other allowed to change. However, this would be laborious and costly.

Professor Udney Yule, professor of statistics Cambridge University UK [2] in 1897, invented the next best method for that purpose, and called it partial regression analysis: the effect of interaction is removed from the data by artificially holding one or more independent variables constant during the analysis. The great merit of partial correlation analysis is not, that it can establish the amount of interaction, but, that it can remove its effects, and, thus, establish what would have happened, if there had been no interaction.

This method of partial regression never received much attention, because it is computationally intensive, and, without the help of a computer, hard to perform, particularly, if you wish to hold *multiple* variables simultaneously constant. Instead, simpler methods are, generally, performed. Even in 1986, Willett and Stampfer [3] assessed the effects of calorie intake and multiple nutrients intake on health by a two stage procedure. First, they adjusted calorie intake for nutrients intake, and, then, they used the adjusted values for assessing the effect on health. This procedure is not adequate, because nutrients intake does not only affect calorie intake but also health, and partial correlation analysis would have been the appropriate approach.

Partial correlation, despite its potential to remove the effect of interactions from the data, is little used so far: searching the Internet we found mainly studies in the field of behavioral sciences like social sciences, marketing, operational research and applied sciences [4]. In clinical research it is virtually unused: when searching Medline, we only found a few time series assessments [5, 6], two psychiatric trials [7, 8], one genetic study [9], and one cardiovascular risk factor study [10]. But this lack of use is probably a matter of time, now that it is available in SPSS (since 2005, SPSS 13) [11] and many other software packages.

This chapter was written as a hand-hold presentation accessible to clinicians, and as a must-read publication for those new to the method. It is the author's experience, as a master class professor, that students are eager to master adequate command of statistical software. For their benefit all of the steps of the novel method from logging in to the final result using SPSS statistical software [11] will be given.

3 Some Theory

Table 5.1 gives the data of a simulated 64 patient study of the effects of exercise on weight loss with calorie intake as covariate. We wish to perform a multiple linear regression of these data with weight loss as dependent (y) and exercise (x_1) and calorie intake (x_2) as independent predictor variables. Because the independent variables should not correlate too strong, first a correlation matrix is calculated (Table 5.2). Correlation coefficients >0.80 or <-0.80 indicate collinearity, and that multiple regression is not valid. This is however, not the case, and we can, thus, proceed. Table 5.3 gives the results of the multiple linear regression. Both calorie intake and exercise are significant independent predictors of weight loss. However, exercise makes you hungry and patients on weight training may be inclined to reduce (or increase) their calorie intake. So, the presence of an interaction between calorie intake and exercise on weight loss is very well possible. In order to check this, an interaction variable ($x_3 =$ calorie intake * exercise, with * symbol of multiplication) is added to the model (Table 5.4). After the addition of the interaction variable to the regression model, exercise is no longer significant and interaction on the outcome is significant at $p=0.002$. There is, obviously, interaction in the study, and the overall analysis of the data is, thus, no longer relevant. The best method to find the true of effect of exercise the study should be repeated with calorie intake

Table 5.1 Data file (Var = variable)

Var 1 weight loss
Var 2 exercise
Var 3 cal intake
Var 4 interaction
Var 5 age

1,00	0,00	1000,00	0,00	45,00
29,00	0,00	1000,00	0,00	53,00
2,00	0,00	3000,00	0,00	64,00
1,00	0,00	3000,00	0,00	64,00
28,00	6,00	3000,00	18000,00	34,00
27,00	6,00	3000,00	18000,00	25,00
30,00	6,00	3000,00	18000,00	34,00
27,00	6,00	1000,00	6000,00	45,00
29,00	0,00	2000,00	000	52,00
31,00	3,00	2000,00	6000,00	59,00
30,00	3,00	1000,00	3000,00	58,00
29,00	3,00	1000,00	3000,00	47,00
27,00	0,00	1000,00	0,00	45,00
28,00	0,00	1000,00	0,00	66,00
27,00	0,00	1000,00	0,00	67,00
28,00	0,00	1000,00	0,00	75,00
2,00	0,00	1000,00	0,00	64,00
30,00	0,00	1000,00	0,00	39,00
1,00	0,00	3000,00	0,00	65,00
3,00	0,00	3000,00	0,00	65,00
25,00	6,00	3000,00	18000,00	52,00
30,00	6,00	3000,00	18000,00	54,00
28,00	6,00	3000,00	18000,00	37,00
29,00	6,00	1000,00	6000,00	52,00
28,00	0,00	2000,00	0,00	65,00
29,00	3,00	2000,00	6000,00	49,00
30,00	3,00	1000,00	3000,00	50,00
30,00	3,00	1000,00	3000,00	51,00
27,00	0,00	1000,00	0,00	40,00
27,00	0,00	1000,00	0,00	46,00
26,00	0,00	1000,00	0,00	59,00
26,00	0,00	1000,00	0,00	53,00
1,00	0,00	1000,00	0,00	42,00
29,00	0,00	1000,00	0,00	53,00
2,00	0,00	3000,00	0,00	47,00
1,00	0,00	3000,00	0,00	54,00
28,00	6,00	3000,00	18000,00	35,00
27,00	6,00	3000,00	18000,00	46,00
30,00	6,00	3000,00	18000,00	56,00
27,00	6,00	1000,00	6000,00	39,00
29,00	0,00	2000,00	0,00	42,00

(continued)

Table 5.1 (continued)

31,00	3,00	2000,00	6000,00	38,00
30,00	3,00	1000,00	3000,00	49,00
29,00	3,00	1000,00	3000,00	50,00
27,00	0,00	1000,00	0,00	51,00
28,00	0,00	1000,00	0,00	64,00
27,00	0,00	1000,00	0,00	65,00
28,00	0,00	1000,00	0,00	59,00
2,00	0,00	1000,00	0,00	53,00
30,00	0,00	1000,00	0,00	72,00
1,00	0,00	3000,00	0,00	65,00
3,00	0,00	3000,00	0,00	47,00
25,00	6,00	3000,00	18000,00	34,00
30,00	6,00	3000,00	18000,00	35,00
28,00	6,00	3000,00	18000,00	34,00
29,00	6,00	1000,00	6000,00	32,00
28,00	0,00	2000,00	0,00	62,00
29,00	3,00	2000,00	6000,00	53,00
30,00	3,00	1000,00	3000,00	47,00
30,00	3,00	1000,00	3000,00	54,00
27,00	0,00	1000,00	0,00	35,00
27,00	0,00	1000,00	0,00	46,00
26,00	0,00	1000,00	0,00	56,00
26,00	0,00	1000,00	45,00	23,00

Table 5.2 Correlation matrix between the variables of the example

Correlations

		Weightloss	Exercise	Calorieintake
Weightloss	Pearson Correlation	1	,405**	−,304*
	Sig. (2-tailed)		,001	,015
	N	64	64	64
Exercise	Pearson Correlation	,405**	1	,390**
	Sig. (2-tailed)	,001		,001
	N	64	64	64
Calorieintake	Pearson correlation	−,304*	,390**	1
	Sig. (2-tailed)	,015	,001	
	N	64	64	64

**Correlation is significant at the 0.01 level (2-tailed)
*Correlation is significant at the 0.05 level (2-tailed)

held constant. Instead of this laborious exercise, a partial correlation analysis with calorie intake held artificially constant can be adequately performed, and would provide virtually the same result. How to do so?

First, the simple linear regression between exercise and weight loss produced a correlation coefficient (r-value) of 0.405, and the multiple correlation coefficient

Table 5.3 Multiple linear regression weight loss as dependent and calorie intake and exercise as independent variables

Coefficients[a]

Model		Unstandardized coefficients		Standardized coefficients		
		B	Std. Error	Beta	t	Sig.
1	(Constant)	29,089	2,241		12,978	,000
	Exercise	2,548	,439	,617	5,802	,000
	Calorieintake	–,006	,001	–,544	–5,116	,000

[a]Dependent variable: weightloss

Table 5.4 Multiple linear regression from previous table with interaction as additional variable

Coefficients[a]

Model		Unstandardized coefficients		Standardized coefficients		
		B	Std. error	Beta	t	Sig.
1	(Constant)	34,279	2,651		12,930	,000
	Interaction	,001	,000	,868	3,183	,002
	Exercise	–,238	,966	–,058	–,246	,807
	Calorieintake	–,009	,002	–,813	–6,240	,000

[a]Dependent variable: weightloss

Table 5.5 Linear correlation of exercise versus weight loss and multiple linear correlation with both calorie intake and exercise versus weight loss

Model summary

Model	R	R square	Adjusted R square	Std. error of the estimate
1	,405[a]	,164	,151	9,73224
1	,644[b]	,415	,396	8,20777

[a]Predictors: (Constant), exercise
[b]Predictors: (Constant), calorieintake, exercise

was larger, 0.644 (Table 5.5). The r-square values are often interpreted as the % certainty about the outcome given by the regression analysis, and we can observe from the example that it rises from 0.164 to 0.415 (Table 5.5), meaning that with two predictors we have 42 instead of 16% certainty. The addition of the second independent variable provided 26% more certainty. This conclusion is correct. But what about the correlation between the second variable (x_2) and the outcome (y). Is it equal to the square root of $0.26 = 0.51$ (51%)? No, because multiple regression is a method that finds the best fit model for all of the data, and, if you add new data, then all of the previously calculated relationships will change.

The change in y caused by the addition of a second variable can be calculated by removing the amount of certainty provided by the presence of a novel variable.

$$\text{Novel y values} = \text{y values} - \text{mean y} - r_{y\,vs\,x2}\left(SD_y\,/\,SD_{x2}\right).$$

SD = standard deviation, vs = versus. Similarly the novel x_1 values can be calculated. Once this has been done for all individuals, an ordinary correlation between the novel values can be calculated. The novel correlation is interpreted as the correlation between y en x_1 with x_2 held constant, and is, otherwise, called the partial correlation.

With multiple x variables, higher order partial correlations can be calculated, which are computationally very intensive, but calculations are pretty much the same. The interpretation is also straightforward, for example, the partial correlation between y en x_1 with two additional x variables is the correlation between y en x_1 with x_2 and x_3 held constant.

What is the clinical relevance of partial correlations. We have come a long way to recognize that, in clinical practice, the outcome of clinical testing is affected by many more factors than a single one, and multiple regression assumes, that all of these factors act independently of one another, but why should that be true [1]. If all of these factors affect the outcome, why should they not affect one another. The point is, that, if you don't have an adequate method to study this issue, you will never know. The great merit of partial correlation is not that it can establish the amount of interaction, but, that it can remove its effects, and, thus, establish what would have happened, if there had not been interaction.

4   Case-Study Analysis

Partial correlation analysis is performed using SPSS, module correlations [11].

Command: Analyze....Correlate....Partial....Variables: enter weight loss and calorie intake....Controlling for: enter exercise....OK.

Table 5.6 upper part shows, that, with exercise held constant, calorie intake is a significant negative predictor of weight loss with a correlation coefficient of −0.548 and a p-value of 0.0001. Also partial correlation with exercise as independent and calorie intake as controlling factor can be performed.

Command: Analyze....Correlate....Partial....Variables: enter weight loss and exercise....Controlling for: enter calorie intake....OK.

Table 5.6 lower part shows that, with calorie intake held constant, exercise is a significant positive predictor of weight loss with a correlation coefficient of 0.596 and a p-value of 0.0001.

It is interesting to observe that the partial correlation coefficient between weight loss and exercise is much larger than the simple correlation coefficient between

Table 5.6 Partial correlation between calorie intake and weight loss with exercise held constant (upper part), partial correlation between exercise and weight loss with calorie intake held constant (lower part)

Correlations				
Control variables			*Weightloss*	*Exercise*
Calorieintake	Weightloss	Correlation	1,000	,596
		Significance (2-tailed)	.	,000
		df	0	61
	Exercise	Correlation	,596	1,000
		Significance (2-tailed)	,000	.
		df	61	0
Control variables			*Weightloss*	*Calorieintake*
Exercise	Weightloss	Correlation	1,000	–,548
		Significance (2-tailed)	.	,000
		df	0	61
	Calorieintake	Correlation	–,548	1,000
		Significance (2-tailed)	,000	.
		df	61	0

Table 5.7 Higher order partial correlation between exercise and weight loss with calorie intake and age held constant

Correlations				
Control variables			Weightloss	Exercise
Age & calorieintake	Weightloss	Correlation	1,000	,541
		Significance (2-tailed)	.	,000
		df	0	60
	Exercise	Correlation	,541	1,000
		Significance (2-tailed)	,000	.
		df	60	0

weight loss and exercise (correlation coefficient = 0.405, Tables 5.2 and 5.5 upper part). Why do we no longer have to account interaction with partial correlations. This is simply because, if you hold a predictor fixed, this fixed predictor can no longer change and interact in a multiple regression model.

Table 5.7 gives an example of higher order partial correlation analysis. Age may affect all of the three variables already in the model. The effect of exercise on weight loss with calorie intake and age fixed is shown. The correlation coefficient is still very significant.

Without the partial correlation approach the conclusion from this study would have been: no definitive conclusion about the effects of exercise and calorie intake is possible, because of a significant interaction between exercise and calorie intake. The partial correlation analysis allows to conclude that both exercise and calorie intake have a very significant linear relationship with weight loss effect.

5   Discussion

The outcome of clinical research is affected by many more factors than a single one. If many factors affect the outcome, why should they not affect one another. Some would even go as far as saying simple linear regression between a factor and an outcome does not exist in real life. E.g. in treatment trials there is always the risk of interaction with factors like age, gender, co-medications, co-morbidities, risk factors etc.

The case study in this chapter shows that a significant interaction on the outcome between subgroup effects makes an overall analysis of the data pretty meaningless, and that in this situation partial correlation analysis can be used to observe the true effect of one variable by artificially keeping the other one constant. How can we clinically interpret interactions like the one between exercise and calorie intake used in the case study. We might say that in the patients with more exercise the calorie intake reduction has better effect, in the patients with less exercise the calorie intake reduction has less effect.

Why not use multiple regression coefficients for adjusting subgroup effects in this situation. The point is that it can only adjust confounders, i.e. independent subgroup effects. In this sense multiple regression is a biased methodology, because subgroups are analyzed as though they work independently of one another although this is virtually never entirely true [1].

Several methods exist for detecting interactions in the data. First, subgroups can be analyzed separately, and, then, compared with one another. Second, analysis of variance is possible. Third, multiple regression with the addition of an interaction variable is possible. The sensitivity of these methods can be somewhat improved by random effect modeling. However, none of these methods are able to establish the amount of interaction or remove its effects from the data.

Partial correlation analysis has a simple principle: if you artificially hold one of the predictors fixed, it can no longer interact with one of the other variables. Partial correlation analysis can neither demonstrate the amount of interaction in the data, but it can remove its effect from the data, and, thus, establish what would have happened, had there been no interaction. We should add that partial correlation analysis is also adequate if used without interaction in the data. However, one will observe that the correlation coefficients will not change after controlling for the additional factors. Nonetheless, it seems also an appropriate method for the *detection* of interaction in your data.

It is time that we added some limitations of partial correlation analysis. First, the calculation of the partial correlation coefficients is based on simple correlation coefficients, assuming linear relationships. It has been observed that interaction effects may not often be strictly linear, and the current software programs are unable to assess non-linear partial correlations. Also as the order of partial correlations goes up, the reliability and sensitivity of the method goes down. However, this is equally true with the standard methods for detecting interactions, particularly with small sample sizes. Third, the calculations are cumbersome and difficult for mathematically un-initiated to understand. However, software has made life a lot easier, and you do not have to be able to perform the calculations in order to understand the methodology. Partial correlation analysis is currently often listed as a machine learning method, because it is computationally intensive and allows computers to remove interaction effects from data files.

6   Conclusions

1. The outcome of clinical research is affected by many more factors than a single one.
2. A significant interaction on the outcome between subgroup effects makes an overall analysis of the data pretty meaningless.
3. Multiple regression is a biased methodology, if subgroup effects interact on the outcome with one another, because it can only adjust confounding but not interaction.
4. Partial correlation analysis can not demonstrate the amount of interaction in the data, but it can remove its effect, and, thus, establish what would have happened, had there been no interaction.

References

1. Cleophas TJ (2003) Sense and nonsense of regression modeling for increasing precision of clinical trials. Clin Pharmacol Ther 74:295–297
2. Yule GU (1897) On the theory of correlation. J R Stat Soc 60:812–854
3. Willett W, Stampfer MJ (1986) Total energy intake: implications for epidemiological analyses. Am J Epidemiol 124:17–22
4. Waliczek TM (1996) A primer on partial correlation coefficients. Governmental Editions, Washington, DC, ED393882
5. Box GEP, Jenkins GM, Reinsel GC (2008) Time series analysis, forecasting and control, 4th edn. Wiley, New York
6. Brockwell P, Davis R (2009) Time series: theory and methods, 2nd edn. Springer, Heidelberg
7. Kazdin AE, French NH, Unis AS, Esveldt-Dawson K, Sherick RB (1983) Hopelessness, depression, and suicidal intent among psychiatrically disturbed inpatient children. J Consult Clin Psychol 51:504–510
8. Mattick RP, Clarke JC (1998) Development and validation of measures of social phobia scrutiny fear and social interaction anxiety. Behav Res Ther 36:455–470
9. Paez JG, Jänne PA, Lee JC et al (2004) EGFR mutations in lung cancer: correlations with clinical response to gefitimib therapy. Science 304:1497–1500
10. Larsson B, Svardsudd K, Welin L, Wilhelmsen L, Bjorntorp P, Tibblin G (1984) Abdominal adipose tissue distribution, obesity, and risk of cardiovascular disease and death: 13 year follow up of participants in the study of men born in 1913. BMJ 288:1401–1409
11. SPSS statistical software www.spss.com. Accessed 08 Oct 2012

Chapter 6
Mixed Linear Models

1 Summary

1.1 Background

In current clinical research repeated measures in a single subject are common. The problem with repeated measures is, that they are more close to one another than unrepeated measures. If this is not taken into account, then data analysis will lose power. In the past decade user-friendly statistical software programs like SAS and SPSS have enabled the application of mixed models as an alternative to the classical general linear model for repeated measures with, sometimes, better sensitivity.

1.2 Objective

This chapter assesses whether in studies with repeated measures, designed to test between-subject differences, the mixed model performs better than does the general linear model.

1.3 Methods and Results

In a parallel group study of cholesterol reducing treatments with five evaluations per patient, the mixed model performed much better than did the general linear model with p-values of respectively 0.0001 and 0.048. In a crossover study of three treatments for sleeplessness the mixed model and general linear model produced similarly well with p-values of 0.005 and 0.010.

T.J. Cleophas and A.H. Zwinderman, *Machine Learning in Medicine*,
DOI 10.1007/978-94-007-5824-7_6, © Springer Science+Business Media Dordrecht 2013

1.4   Conclusions

We conclude that mixed models do, indeed, seem to produce better sensitivity of testing, when there are small within-subject differences and large-between subject-differences, and when the main objective of your research is to demonstrate between- rather than within-subject differences.

The novel mixed model may be more complex. Yet, with modern user-friendly statistical software its use is straightforward, and its software-commands are no more complex than they are with standard methods. We hope that this chapter will encourage clinical researchers to more often make use of its benefits.

2   Introduction

In current clinical research repeated measures in a single subject are common. The problem with repeated measures is that they are more close to one another than unrepeated measures. If this is not taken into account, then data analysis will lose power. The classical general linear analysis of variance (ANOVA) model for the analysis of such data has been recently supplemented by a novel method, mixed linear modeling. Mixed modeling was first described by Henderson [1] in the early 1960s as a linear model for making predictions from longitudinal observations in a single subject. It first became popular in the early 1990s by the work of Robinson [2] and McLean [3] who improved the model by presenting consistent analysis procedures. In the past decade user-friendly statistical software programs like SAS [4] and SPSS [5] have enabled the application of mixed models even by clinical investigators with limited statistical background.

With mixed models repeated measures *within* subjects receive fewer degrees of freedom than they do with the classical general linear model, because they are nested in a separate layer or subspace. In this way better sensitivity is left in the model to demonstrate differences *between* subjects. Therefore, if the main aim of your research is to demonstrate differences *between* subjects, then the mixed model should be more sensitive. However, the two methods should be equivalent if the main aim of your research is to demonstrate differences between repeated measures, for example different treatment modalities in a single subject. A limitation of the mixed model is that it includes additional variances, and is, therefore, more complex. More complex statistical models are, ipso facto, more at risk of power loss, particularly, with small data.

The current chapter uses examples to demonstrate how the mixed model performs in practice. The examples show that the mixed model, unlike the general linear, produced a very significant effect in a parallel-group study with repeated measures, and that the two models were approximately equivalent for analyzing a crossover study with three treatment modalities. We hope this explanatory chapter will be helpful to researchers assessing repeated measures.

Table 6.1 Two parallel groups of ten patients are assessed five times for their HDL-cholesterol

Patient no	Week 1	Week 2	Week 3	Week 4	Week 5	Treatment modality
1	1,66	1,62	1,57	1,52	1,50	0,00
2	1,69	1,71	1,60	1,55	1,56	0,00
3	1,92	1,94	1,83	1,78	1,79	0,00
4	1,95	1,97	1,86	1,81	1,82	0,00
5	1,98	2,00	1,89	1,84	1,85	0,00
6	2,01	2,03	1,92	1,87	1,88	0,00
7	2,04	2,06	1,95	1,90	1,91	0,00
8	2,07	2,09	1,98	1,93	1,94	0,00
9	2,30	2,32	2,21	2,16	2,17	0,00
10	2,36	2,35	2,26	2,23	2,20	0,00
11	1,57	1,82	1,83	1,83	1,82	1,00
12	1,60	1,85	1,89	1,89	1,85	1,00
13	1,83	2,08	2,12	2,12	2,08	1,00
14	1,86	2,11	2,16	2,15	2,11	1,00
15	2,80	2,14	2,19	2,18	2,14	1,00
16	1,92	2,17	2,22	2,21	2,17	1,00
17	1,95	2,20	2,25	2,24	2,20	1,00
18	1,98	2,23	2,28	2,27	2,24	1,00
19	2,21	2,46	2,57	2,51	2,48	1,00
20	2,34	2,51	2,55	2,55	2,52	1,00

Treatment modality 0 = placebo, 1 = active treatment
SPSS uses commas instead of dots

3 A Placebo-Controlled Parallel Group Study of Cholesterol Treatment

A placebo-controlled parallel group study cholesterol is used as first example. Each patient is measured for HDL-cholesterol level for 5 weeks once a week. Table 6.1 gives the data file. A significant difference between the two treatments is expected. The graph of the summaries of the data (Fig. 6.1) shows, indeed, that already after 2 weeks the treatment-1 group (active treatment group) starts to perform better than the treatment-0 group (placebo-group). Multiple unpaired t-tests of treatment-0 versus treatment-1 demonstrate significant differences with p-values as small as 0.08 at the weeks 3, 4 and 5 (analysis not shown). However, this analysis is not entirely appropriate, because it does not take the repeated nature of the data into account. A repeated measurements analysis of variance using the classical general linear model is performed (full factorial design). SPSS statistical software is used [4].

We command
Analyze....General linear model....Repeated Measurements....Define factors....
Within-subjects factor names: week....number levels: 5....add....Define.... enter
week 1, 2, 3, 4, 5 in box: "Within-subjects Variables"....enter treatment in box between-subjects covariates....OK.

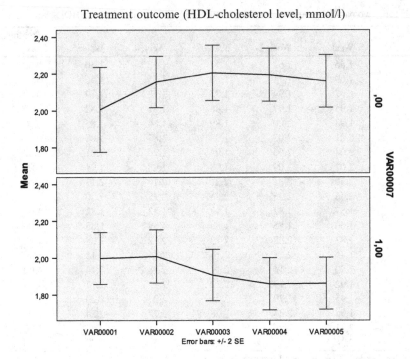

Treatment outcome (HDL-cholesterol level, mmol/l)

Fig. 6.1 Two parallel groups treated for hypercholesterolemia for 5 weeks (0.00 = placebo treatment; 1.00 = active treatment)

The Table 6.2 gives the results. As the test for sphericity (equal standard errors) had to be rejected (not shown), sphericity could not be assumed, and the Huynh-Feldt test was the next best for demonstrating a difference between the repeated measures (Table 6.2: source week). With p=0.115 no significant difference between the repeated measures could be demonstrated (upper part of Table 6.2). The subsequent between-subjects comparison of the two treatments showed a borderline effect with p-value of 0.048 (lower part of Table 6.2).

As an alternative a mixed linear model is applied using the same version of SPSS. For that purpose the data file has to be adapted. Every week must be given a separate row (Table 6.3).

We command
Analyze....mixed models....linear....specify subjects and repeated....variable 1 ....continue....linear mixed model....dependent: variable 3....factors: variable 2, variable 4....fixed....build nested term....variable 4....add....variable 2....add.... variable 2 build term by* variable 4....variable 4 * variable 2....add.... continue.... OK (* = sign of multiplication).

Table 6.4 gives the results. With the mixed model analysis the treatment modality has become a very significant predictor of treatment outcome with p < 0.0001.

Table 6.2 Repeated measures ANOVA using the classical general linear model for the analysis of the data from Table 6.1

Tests of within-subjects effects

Measure: measure 1

Source		Type III sum of squares	df	Mean square	F	Sig.
Week	Sphericity Assumed	,089	4	,022	2,692	,038
	Greenhouse-Geisser	,089	1,022	,087	2,692	,117
	Huynh-Feldt	,089	1,086	,082	2,692	,115
	Lower-bound	,089	1,000	,089	2,692	,118
Week* treatment	Sphericity Assumed	,380	4	,095	11,460	,000
	Greenhouse-Geisser	,380	1,022	,372	11,460	,003
	Huynh-Feldt	,380	1,086	,350	11,460	,003
	Lower-bound	,380	1,000	,380	11,460	,003
Error(week)	Sphericity Assumed	,597	72	,008		
	Greenhouse-Geisser	,597	18,396	,032		
	Huynh-Feldt	,597	19,550	,031		
	Lower-bound	,597	18,000	,033		

Tests of between-subjects effects

Measure: measure_1

Transformed variable :average

Source	Type III sum of squares	df	Mean square	F	Sig.
Intercept	414,530	1	414,530	1573,798	,000
Treatment	1,188	1	1,188	4,511	,048
Error	4,741	18	,263		

As the test for sphericity was rejected (not shown), sphericity could not be assumed, and the Huynh-Feldt test was the next best for demonstrating a difference between the repeated measures (source: week). With $p = 0.115$ no significant difference between the repeated measures was demonstrated (upper table). The subsequent between-subjects comparison showed a borderline effect with p-value of 0.048 (lower table)

This result is in agreement with our prior expectation, and it is also more similar to the above unpaired t-tests than the general linear model is.

We can now conclude that, after adjustment for the repeated nature of the data, there is a very significant difference between the effect of the placebo and the active treatment on HDL cholesterol a $p < 0.0001$.

4   A Three Treatment Crossover Study of the Effect of Sleeping Pills on Hours of Sleep

A three treatment crossover study of the effect of sleeping pills on hours of sleep is used as a second example. Table 6.5 gives the data file. Ten patients are given three different sleeping pills in a randomized double-blind fashion. The hours of sleep

Table 6.3 The data from Table 6.1 adapted for mixed linear modeling, each patient is now included five times (five rows), (SPSS uses commas instead of dots)

1#	2	3	4	1	2	3	4
Patient	Week	Outcome	Treatment	Patient	Week	Outcome	Treatment
1,00	1,00	1,66	,00	10,00	3,00	2,26	0,00
1,00	2,00	1,62	,00	10,00	4,00	2,23	0,00
1,00	3,00	1,57	,00	10,00	5,00	2,20	0,00
1,00	4,00	1,52	,00	11,00	1,00	1,57	1,00
1,00	5,00	1,50	,00	11,00	2,00	1,82	1,00
2,00	1,00	1,69	,00	11,00	3,00	1,83	1,00
2,00	2,00	1,71	,00	11,00	4,00	1,83	1,00
2,00	3,00	1,60	,00	11,00	5,00	1,82	1,00
2,00	4,00	1,55	,00	12,00	1,00	1,60	1,00
2,00	5,00	1,56	,00	12,00	2,00	1,85	1,00
3,00	1,00	1,92	,00	12,00	3,00	1,89	1,00
3,00	2,00	1,94	,00	12,00	4,00	1,89	1,00
3,00	3,00	1,83	,00	12,00	5,00	1,85	1,00
3,00	4,00	1,78	,00	13,00	1,00	1,83	1,00
3,00	5,00	1,79	,00	13,00	2,00	2,08	1,00
4,00	1,00	1,95	,00	13,00	3,00	2,12	1,00
4,00	2,00	1,97	,00	13,00	4,00	2,12	1,00
4,00	3,00	1,86	,00	13,00	5,00	2,08	1,00
4,00	4,00	1,81	,00	14,00	1,00	1,86	1,00
4,00	5,00	1,82	,00	14,00	2,00	2,11	1,00
5,00	1,00	1,98	,00	14,00	3,00	2,16	1,00
5,00	2,00	2,00	,00	14,00	4,00	2,15	1,00
5,00	3,00	1,89	,00	14,00	5,00	2,11	1,00
5,00	4,00	1,84	,00	15,00	1,00	2,80	1,00
5,00	5,00	1,85	,00	15,00	2,00	2,14	1,00
6,00	1,00	2,01	,00	15,00	3,00	2,19	1,00
6,00	2,00	2,03	,00	15,00	4,00	2,18	1,00
6,00	3,00	1,92	,00	15,00	5,00	2,14	1,00
6,00	4,00	1,87	,00	16,00	1,00	1,92	1,00
6,00	5,00	1,88	,00	16,00	2,00	2,17	1,00
7,00	1,00	2,04	,00	16,00	3,00	2,22	1,00
7,00	2,00	2,06	,00	16,00	4,00	2,21	1,00
7,00	3,00	1,95	,00	16,00	5,00	2,17	1,00
7,00	4,00	1,90	,00	17,00	1,00	1,95	1,00
7,00	5,00	1,91	,00	17,00	2,00	2,20	1,00
8,00	1,00	2,07	,00	17,00	3,00	2,25	1,00
8,00	2,00	2,09	,00	17,00	4,00	2,24	1,00
8,00	3,00	1,98	,00	17,00	5,00	2,20	1,00
8,00	4,00	1,93	,00	18,00	1,00	1,98	1,00
8,00	5,00	1,94	,00	18,00	2,00	2,23	1,00
9,00	1,00	2,30	,00	18,00	3,00	2,28	1,00
9,00	2,00	2,32	,00	18,00	4,00	2,27	1,00

(continued)

Table 6.3 (continued)

1#	2	3	4	1	2	3	4
Patient	Week	Outcome	Treatment	Patient	Week	Outcome	Treatment
9,00	3,00	2,21	,00	18,00	5,00	2,24	1,00
9,00	4,00	2,16	,00	19,00	1,00	2,21	1,00
9,00	5,00	2,17	,00	19,00	2,00	2,46	1,00
10,00	1,00	2,36	,00	19,00	3,00	2,57	1,00
10,00	2,00	2,35	,00	19,00	4,00	2,51	1,00
				19,00	5,00	2,48	1,00
				20,00	1,00	2,34	1,00
				20,00	2,00	2,51	1,00
				20,00	3,00	2,55	1,00
				20,00	4,00	2,55	1,00
				20,00	5,00	2,52	1,00

[#]1 = patient number, 2 = week of treatment (1–5), 3 = outcome (HDL cholesterol), 4 = treatment modality (0 or 1)

Table 6.4 Mixed model analysis of the data from Table 6.2 with treatment modality and week of treatment as predictors, and treatment outcome as dependent variable

Type III tests of fixed effects[a]

Source	Numerator df	Denominator df	F	Sig.
Intercept	1	76,570	6988,626	,000
Week	4	31,149	,384	,818
Treatment	1	76,570	20,030	,000
Week * treatment	4	31,149	1,337	,278

The treatment modality is a very significant predictor of treatment outcome
[a]Dependent variable: outcome

Table 6.5 A single group is assessed for three different treatments (hours of sleep) in a crossover fashion, age and gender are in the columns 4 and 5 (SPSS uses commas instead of dots)

Patient no	Treatment 1	Treatment2	Treatment 3	Age	Gender
1	6,10	6,80	5,20	55,00	0,00
2	7,00	7,00	7,90	65,00	0,00
3	8,20	9,00	3,90	74,00	0,00
4	7,60	7,80	4,70	56,00	1,00
5	6,50	6,60	5,30	44,00	1,00
6	8,40	8,00	5,40	49,00	1,00
7	6,90	7,30	4,20	53,00	0,00
8	6,70	7,00	6,10	76,00	0,00
9	7,40	7,50	3,80	67,00	1,00
10	5,80	5,80	6,30	66,00	1,00

Treatment outcome (hours of sleep/night)

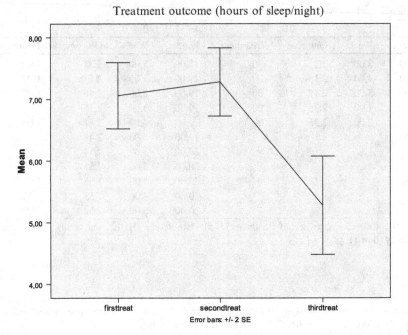

Fig. 6.2 Crossover study of the effect on hours of sleep of three sleeping pills in a single group of patients

after treatment are the outcome variable. A significant difference between the three treatment outcomes is expected. Figure 6.2 is a graph of the summaries of the treatment effects. The third treatment seems to perform significantly worse than the other two treatments as indicated by the error bars giving 95% confidence intervals of the means. However, this conclusion is not entirely appropriate, because it does not take the repeated nature of the data into account.

We command
Analyze....General linear model....Repeated Measurements....Define factors.... Within-subjects factor names: treatment....number levels: 3....add....Define.... enter treatment 1, 2, 3 in box: "Within-subjects Variables"....enter gender in box between-subjects covariates....OK.

Table 6.6 gives the results. As the test for sphericity (equal standard errors) had to be rejected again (not shown), sphericity could not be assumed, and the Huynh-Feldt test was the next best for demonstrating a difference between the repeated measures (Table 6.6: source treatment). With p=0.010 a very significant difference between the repeated measures, the treatments, could be demonstrated (upper part of Table 6.2). The subsequent between-subjects comparison of the two genders did not produce a significantly different effect of the genders on the treatment outcome with a p-value of 0.65 (lower part of Table 6.6).

Table 6.6 Repeated measures ANOVA using the classical general linear model for the analysis of the data from Table 6.5

Tests of within-subjects effects

Measure: MEASURE 1

Source		Type III sum of squares	df	Mean square	F	Sig.
Treatment	Sphericity assumed	24,056	2	12,028	9,642	,002
	Greenhouse-geisser	24,056	1,024	23,494	9,642	,014
	Huynh-feldt	24,056	1,181	20,369	9,642	,010
	Lower-bound	24,056	1,000	24,056	9,642	,015
Treatment * gender	Sphericity assumed	,392	2	,196	,157	,856
	Greenhouse-geisser	,392	1,024	,383	,157	,708
	Huynh-feldt	,392	1,181	,332	,157	,742
	Lower-bound	,392	1,000	,392	,157	,702
Error(treatment)	Sphericity assumed	19,959	16	1,247		
	Greenhouse-geisser	19,959	8,191	2,437		
	Huynh-feldt	19,959	9,448	2,112		
	Lower-bound	19,959	8,000	2,495		

Tests of between-subjects effects

Measure: MEASURE_1

Transformed variable: Average

Source	Type III sum of squares	df	Mean square	F	Sig.
Intercept	1283,148	1	1283,148	1472,063	,000
Gender	,192	1	,192	,220	,651
Error	6,973	8	,872		

As the test for sphericity was rejected (not shown), sphericity could not be assumed, and the Huynh-Feldt test was the next best for demonstrating a difference between the repeated measures (source: treat number). A significant difference between the repeated measures was demonstrated with p=0.010 (upper table). The subsequent between-subjects comparison did not show a significant effect of gender with p-value of 0.65 (lower table)

As an alternative a mixed linear model is applied using the same version of SPSS. For that purpose the data file has to be adapted. Each treatment (3) must be given a separate row. Table 6.7 gives the adapted data file made adequate for mixed linear modeling.

We command
Analyze….mixed models….linear….specify subjects and repeated….variable 1…. Continue….linear mixed model….dependent: variable 3….factors: variable 2, variable 4….fixed….build nested term….variable 4….add….variable 2….add …. variable 2 build term by* variable 4….variable 4 * variable 2….add…. continue….OK (*=sign of multiplication).

Table 6.8 gives the results. With the mixed model analysis the treatment number is a very significant predictor of treatment outcome with p<0.005. This result is in

Table 6.7 The data from Table 6.5 adapted for mixed linear modeling, each patient is included three times (three rows), (SPSS uses commas instead of dots)

Variable			
1	2	3	4
1,00	1,00	6,10	0,00
1,00	2,00	6,80	0,00
1,00	3,00	5,20	0,00
2,00	1,00	7,00	0,00
2,00	2,00	7,00	0,00
2,00	3,00	7,90	0,00
3,00	1,00	8,20	0,00
3,00	2,00	9,00	0,00
3,00	3,00	3,90	0,00
4,00	1,00	7,60	1,00
4,00	2,00	7,80	1,00
4,00	3,00	4,70	1,00
5,00	1,00	6,50	1,00
5,00	2,00	6,60	1,00
5,00	3,00	5,30	1,00
6,00	1,00	8,40	1,00
6,00	2,00	8,00	1,00
6,00	3,00	5,40	1,00
7,00	1,00	6,90	0,00
7,00	2,00	7,30	0,00
7,00	3,00	4,20	0,00
8,00	1,00	6,70	0,00
8,00	2,00	7,00	0,00
8,00	3,00	6,10	0,00
9,00	1,00	7,40	1,00
9,00	2,00	7,50	1,00
9,00	3,00	3,80	1,00
10,00	1,00	5,80	1,00
10,00	2,00	5,80	1,00
10,00	3,00	6,30	1,00

1 = patient number, 2 = treatment number, 3 = outcome, 4 = gender

Table 6.8 Mixed model analysis of the data from Table 6.6 with treatment number as predictor, and treatment outcome as dependent variable

Type III tests of fixed effects[a]				
Source	Numerator df	Denominator df	F	Sig.
Intercept	1	17,574	1143,456	,000
Gender	1	17,574	,171	,684
Treatmentnumber	2	11,628	8,470	,005
Treatmentnumber * gender	2	11,628	,202	,820

The treatment number is a very significant predictor of the treatment outcome
[a]Dependent variable: outcome

agreement with our prior expectation, and it is also rather similar to the above general linear model analysis (p=0.01).

We can now conclude that after adjustment for the repeated nature of the data there is a significantly different effect of the three treatment modalities on the hours of sleep with a p-value between 0.005 and 0.01.

5   Discussion

The examples given in the current report confirm the primary hypothesis of this chapter. In the first example the aim of the research was to demonstrate a difference between the two treatment modalities given to two parallel groups of patients, while the differences between the repeated measures in a single patient was less important. Indeed, the mixed model produced a much better result than did the general linear model with overall p-values of <0.0001 versus only 0.048 for treatment effect. In the second example the main aim was to demonstrate a significant difference between the repeated measures in a single patient. The general linear model produced a significant p-value at 0.010, compared to a much similar p-value of 0.005 in the mixed model. Mixed models do, indeed, seem to produce better sensitivity of testing, when there are small within-subject differences and large-between subject-differences.

As indicated in the introduction, a disadvantages of the novel model is, that it is more complex, and, therefore, may require a larger sample size than the general linear model. Yet, as demonstrated in the examples, it may perform well even with samples as small as 10–20, that is, if the model being used is not too complex.

Apart from the advantage that it can sometimes handle between-subject differences with more sensitivity, it may provide a number of additional advantages including:

1. the general linear model will drop cases with missing data entirely, whereas the mixed model can include incomplete cases in the analysis;
2. the general linear model assumes that all subjects are measured at the same point of time, whereas the mixed model allows subjects to be measured at different points of time;
3. the general linear model requires subjects to have equal numbers of repeated measurements, whereas the mixed model allows unequal repetitions;
4. the presence of sphericity (equal standard errors) is not a requirement of the mixed models.

We conclude that the mixed model is a welcome supplement to the commonly-used general linear models for the overall analysis of studies with repeated measures. Unfortunately, although widely discussed in methodology papers [6–10], the mixed model is little used by clinical researchers in practice so far. This may be due to its obvious complexity. Yet, the current paper shows that with modern user-friendly statistical software its use is straightforward, and its software commands are no more complex than they are with standard methods for data analysis. We do hope that this chapter will encourage clinical researchers to more often make use of the benefits of the mixed model.

6   Conclusion

In current clinical research repeated measures in a single subject are common. The problem with repeated measures is, that they are more close to one another than unrepeated measures. If this is not taken into account, then data analysis will lose power. In the past decade user-friendly statistical software programs like SAS and SPSS have enabled the application of mixed models as an alternative to the classical general linear model for repeated measures with, sometimes, better sensitivity. This chapter assesses whether in studies with repeated measures, designed to test between-subject differences, the mixed model performs better than does the general linear model.

In a parallel group study of cholesterol reducing treatments with five evaluations per patient, the mixed model performed much better than did the general linear model with p-values of respectively 0.0001 and 0.048. In a crossover study of three treatments for sleeplessness the mixed model and general linear model produced similarly well with p-values of 0.005 and 0.010.

We conclude that mixed models do, indeed, seem to produce better sensitivity of testing, when there are small within-subject differences and large-between subject-differences, and when the main objective of your research is to demonstrate between- rather than within-subject differences.

The novel mixed model may be more complex. Yet, with modern user-friendly statistical software its use is straightforward, and its software-commands are no more complex than they are with standard methods. We hope that this chapter will encourage clinical researchers to more often make use of its benefits.

References

1. Henderson CR, Kempthorne O, Searle SR, Von Krosigk CM (1959) The estimation of environmental and genetic trends from records subject to culling. Biometrics 15:192–218
2. Robinson GK (1991) That best linear unbiased predictions (BLUP) is a good thing. Stat Sci 6:15–32
3. McLean RA, Sanders WL, Stoup WW (1991) A unified approach to mixed linear models. Am Stat 45:54–64
4. SAS statistical software (2010) www.sas.com. Accessed 06 July 2010
5. SPSS statistical software 18.0. (2010) www.spss.com. Accessed 06 July 2010
6. Milliken GA, Johnson DE (1992) Analysis of messy data. Designed experiments. Chapman and Hall, New York

7. West BT, Welch KB, Galecki AT (2007) Linear mixed models: a practical guide to using statistical software. Chapman and Hall, New York
8. Breslow NE, Clayton DG (1993) Approximate inferences in generalized linear mixed models. J Am Stat Assoc 88:9–25
9. Engel B, Keen A (1994) A simple approach for the analysis of generalized linear mixed models. Stat Neerlandica 48:1–22
10. Littell RC, Milliken GA, Stroup WW, Wolfinger RD (1996) SAS system for mixed models. SAS Institute Inc., Cary

Chapter 7
Binary Partitioning

1 Summary

1.1 Background

Binary partitioning can assist physicians in diagnosing patients potentially suffering heart attacks and other clinical conditions. Traditionally, the physicians made decisions based on their clinical experience. Classifications based on representative historical data has the advantage of added empirical information from large numbers of patients.

1.2 Objective

This chapter is to familiarize the research community with this important methodology for improving the diagnostic accuracy of prognostic clinical decision analysis.

1.3 Methods and Results

An example is used to explain The ROC (receiver operating characteristic) and entropy methods for simple partitions.

ROC curves are used for finding the best cut-off levels in a decision tree of diagnostic procedures. The problem with ROC curves is that the sample sizes of the positive and negative tests are not taken into account, reducing the power of this approach. This is, particularly, a problem if the numbers of positive and negative tests are largely different in size. The entropy method based on log likelihood statistics is helpful in this situation. The theory sounds rather complex, but the equations work smoothly, and can be performed even on a pocket calculator.

1.4   Conclusions

We conclude that binary partition and its closely related decision trees may become one of the standard analytic choices in clinical research, but they likely complement rather than replace the classic statistical methods. For assessing the level of statistical significance of assumed differences and effects the classic statistical methods are more suitable.

2   Introduction

One of the most important and original applications of binary partitioning was to develop data-based decision cut-off levels that can assist physicians in diagnosing patients potentially suffering heart attacks [1]. Traditionally, the physicians made decisions based on their clinical experience. Also, laboratories developed diagnostic tests with normal values based on rather intuitive grounds. Classifications based on representative historical data has the advantage of added empirical information from large numbers of patients. This is, particularly, important if symptoms, signs and diagnostic procedures give rise to a substantial number of false positive and false negative results as often observed in clinical practice. The main purpose of the data-based methods is to reduce the latter number. The book by Breiman et al. [2] on classification and regression trees is a milestone on binary partitioning and closely related cut-off decision trees, otherwise called CART (classification and regression) trees. The associated CART program has become a commercial software [3], but simple partitions and decision trees can also be performed on a pocket calculator. This chapter was written to familiarize the research community with this important methodology for improving the diagnostic accuracy of clinical decision trees.

3   Example

Several vascular labs have defined their estimators for peripheral vascular disease as shown underneath [4].

1. Ankle blood pressure	>brachial blood pressure
2. Ankle pressure after 5 min treadmill	>20% reduction from baseline
3. Proximal thigh pressure	>30 mmHg above brachial pressure
4. Segmental pressures (thigh, calf, ankle)	<20 mmHg difference between two levels
5. Toe pressure	<40% different from brachial pressure.

Considering the rounded pattern of the above "normal" values for predicting the presence of vascular disease, we may assume that the values as given are based on empiricism and agreement rather than calculated averages, but we can do better.

Although the accumulated evidence of the assessment may give rise to a sensitivity close to 95% and a specificity close to 95%, both sensitivity and specificity may increase to 98 or 99% if we fine-tune the cut-off levels of the contributory estimators using representative historical data.

For that purpose a representative sample of patients has to be assessed against a golden standard, i.e., angiography in the given example. The entire sample can be split into patients with a higher and those with an equal or lower ankle blood pressure than their brachial blood pressure, because we know that this is a major symptom of vascular disease. The procedure of splitting is less straightforward, if the estimators are quantitative and multiple cut-off levels are possible like the above no. 2–5 estimators show.

4 ROC (Receiver Operating Characteristic) Method for Finding the Best Cut-off Level

A hypothesized example of a cut-off level is given in Fig. 7.1. If a fall of ankle blood pressure after 5 min treadmill exercise of > 26% is used as threshold for a positive test, then the number of patients with a true positive test are "a", true negative "b", false positive "c", and false negative "d". The ratio "a/(a+d)" is called the sensitivity of the test, the ratio "b/(b+c)" the specificity of the test.

Underneath, an overview is given of the calculated sensitivities and specificities if different cut-off levels are applied.

Cut-off level	Sensitivity	Specificity
22%	1.000	0.723
23%	0.997	0.701
24%	0.990	0.855
25%	0.980	0.908
26%	**0.960**	**0.950**
27%	0.940	0.972
28%	0.910	0.986
29%	0.860	0.993
30%	0.800	1.000

We would like to have a sensitivity and specificity close to 100%. However, in practice most diagnostic tests are far from perfect and produce false positive and false negative effects. With qualitative tests there is little we can do. With quantitative tests we can increase the sensitivity by moving the vertical decision line between a positive and negative test to the left (Fig. 7.1), and we can increase specificity by moving it in the opposite direction. Figure 7.2 shows the relationship between sensitivities and specificities. The curve suggests a high sensitivity and at the same time high specificity, if 26% or 27% are used as cut-off levels for a positive test. The best cut-off level is obtained if the distance from the curve at that point to the top of the

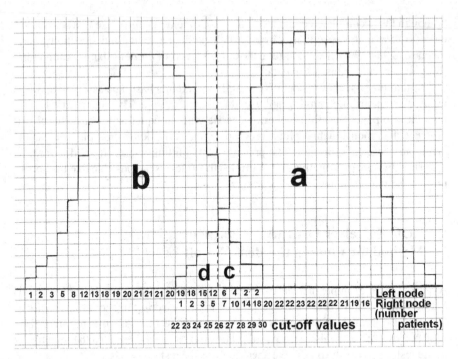

Fig. 7.1 Histogram of a patients' sample assessed for peripheral vascular disease; "*a*" summarizes the patients with a positive test and the presence of disease, "*b*" the patients with a negative test and the absence of disease, "*c*" and "*d*" are the false positive and false negative patients respectively, if 26 is used as a cut-off value between a positive and a negative test

y-axis is closest. The 26% cut-off level can be calculated to provide the closest distance: Pythagoras's equation for right angular triangles shows a distance of $\sqrt{(4^2+5^2)}=\sqrt{41}$ which is closer than the shortest distance next to it $(\sqrt{(6^2+3^2)}=\sqrt{45})$. This result is, thus, a better predictor for vascular disease than the value of 20% as previously agreed in the vascular laboratory on intuitive grounds [4]. In this manner the ROC method can be used for determining the best cut-off level to be included in a decision tree of diagnostic procedures like the above one.

The problem with ROC method is that the sample sizes of the positive and negative tests are not taken into account, reducing the power of this approach. This is, particularly, a problem if the numbers of positive and negative tests are largely different in size. The entropy method is helpful in this situation.

5   Entropy Method for Finding the Best Cut-off Level

The entropy method has an interesting history. It received its name, because it makes use of an equation that was formerly applied in science to estimate the amount of energy loss in thermodynamics, but, otherwise, has no connection with its application

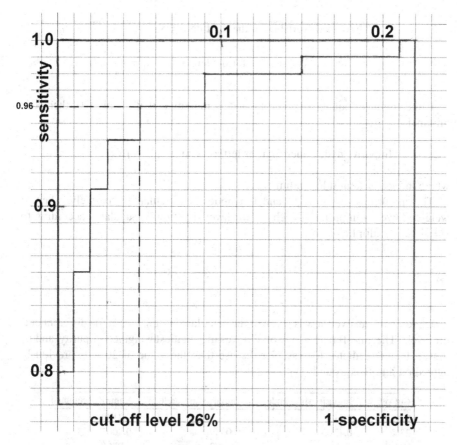

Fig. 7.2 ROC curve of threshold values of positive tests for vascular disease based on the % ankle blood pressure reduction after 5 min treadmill. The shortest distance to the top of the y-axis, and, thus, the best predictive test is obtained with 26% blood pressure reduction as cut-off value

to science. A pleasant thing about this method is that, unlike the ROC method, the result can be easily adjusted for magnitude of the samples. This is also the reason that the result of the ROC method often slightly differs from that of the entropy method.

In entropy-method-terminology the entire sample of patients (Fig. 7.1) is called the parent node, which can, subsequently, be repeatedly split, partitioned if you will, into binary internal nodes. Mostly, internal nodes contain false positive or negative patients, and are, thus, somewhat impure. The magnitude of their impurity is assessed by the log likelihood method [5]. Impurity equals the maximum log likelihood of the y-axis-variable by assuming that the x-axis-variable follows a Gaussian (i.e. binomial) distribution and is expressed in units, sometimes called bits (a short-cut for "binary digits"). All this sounds rather complex, but it works smoothly.

$$\text{The x-axis variable for the right node} = x_r = \frac{a}{a+d},$$

$$\text{for the left node} = x_l = \frac{b}{b+c}.$$

If the impurity equals 1.0 bits, then it is maximal, if it equals 0.0, then it is minimal.

$$\text{Impurity node either right or left} = -x \ln x - (1-x)\ln(1-x),$$

where in means natural logarithm.

The impurities of the right and left node are calculated separately. Then, a weighted overall impurity of each cut-off level situation is calculated according to (*=sign of multiplication):

Underneath, an overview is given of the calculated impurities at the different cut-off levels. The cut-off percentage of 27 gives the smallest weighted impurity, and is, thus, a better predictor for vascular disease than the value of 20% as previously agreed on intuitive grounds [1].

Cut-off	Impurity right node	Impurity left node	Impurity weighted
22%	0.5137	0.0000	0.3180
23%	0.4392	0.0559	0.3063
24%	0.4053	0.0982	0.2766
25%	0.3468	0.1352	0.2711
26%	0.1988	0.1688	0.1897
27%	**0.1352**	**0.2268**	**0.1830**
28%	0.0559	0.3025	0.1850
29%	0.0559	0.3850	0.2375
30%	0.0000	0.4690	0.2748

Also, it should be a better predictor of vascular disease than the 26% value as established by the ROC method, because, unlike the ROC method, the entropy method takes into account and adjusts the differences in sample sizes of the nodes.

6   Discussion

The best cut-off level for making optimal predictions from diagnostic tests can be calculated from the ROC and entropy methods. The next step is, of course, to include multiple tests in order to further increase accuracy of making predictions.

Decision trees can help making rapid clinical decisions, and these decisions are more accurate if they are based on calculation instead of intuition. The cut-off levels used in the example of this chapter could, after the decision tree procedure, look somewhat like the levels shown below with accompanying sensitivities/specificities of 99% instead of 95%.

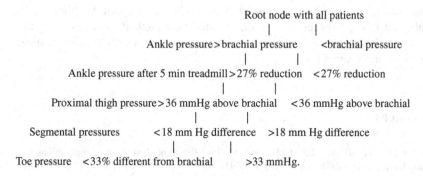

Root node with all patients

Ankle pressure > brachial pressure   < brachial pressure

Ankle pressure after 5 min treadmill > 27% reduction   < 27% reduction

Proximal thigh pressure > 36 mmHg above brachial   < 36 mmHg above brachial

Segmental pressures   < 18 mm Hg difference   > 18 mm Hg difference

Toe pressure   < 33% different from brachial   > 33 mmHg.

We should add that, in addition to mathematical arguments, there may be clinical arguments for setting the cut-off levels for sensitivity and specificity. For example, for incurable deadly diseases, you may want to avoid false positives, meaning telling a healthy person he/she will die soon, while false negatives are not so bad since you can not treat the condition anyway. If, instead, the test would serve as a first screening for a fatal condition if untreated but completely treatable, it should provide a better sensitivity even at the expense of a lower specificity.

We conclude that the binary partition and its closely related decision trees may become one of the standard analytic choices in clinical disease, but they likely complement rather than replace the classic statistical methods. For assessing the level of statistical significance of assumed differences and effects the classic statistical methods are more suitable.

7 Conclusions

Binary partitioning can assist physicians in diagnosing patients potentially suffering heart attacks and other clinical conditions. Traditionally, the physicians made decisions based on their clinical experience. Classifications based on representative historical data has the advantage of added empirical information from large numbers of patients. This chapter is to familiarize the research community with this important methodology for improving the diagnostic accuracy of prognostic clinical decision analysis.

An example is used to explain The ROC (receiver operating characteristic) and entropy methods for simple partitions.

ROC curves are used for finding the best cut-off levels in a decision tree of diagnostic procedures. The problem with ROC curves is that the sample sizes of the positive and negative tests are not taken into account, reducing the power of this approach. This is, particularly, a problem if the numbers of positive and negative tests are largely different in size. The entropy method based on log likelihood statistics is helpful in this situation. The theory sounds rather complex, but the equations work smoothly, and can be performed even on a pocket calculator.

We conclude that binary partition and its closely related decision trees may become one of the standard analytic choices in clinical research, but they likely complement rather than replace the classic statistical methods. For assessing the level of statistical significance of assumed differences and effects the classic statistical methods are more suitable.

References

1. Wasson JH, Sox HC, Neff RK, Goldman L (1985) Clinical prediction rules: applications and methodologic standards. N Engl J Med 313:793–799
2. Breiman L, Friedman JH, Olshen RA, Stone CJ (1984) Classification and regression trees. Chapman & Hall (Wadsworth Inc.), New York
3. Salford Predictive Modeler. www.salford-systems.com
4. Jaff MR, Dorros G (1998) The vascular laboratory: a critical component required for successful management of peripheral arterial occlusive disease. J Endovasc Surg 5:146–158
5. Cleophas TJ, Zwinderman AH, Van Ouwerkerk BM (2007) Log likelihood ratio tests for the assessment of cardiovascular events. Perfusion 20:79–82

Chapter 8
Item Response Modeling

1 Summary

1.1 Background

Item response models using exponential modeling are more sensitive than classical linear methods for making predictions from psychological questionnaires.

1.2 Objective

This chapter is to assess whether they can also be used for making predictions from quality of life questionnaires and clinical and laboratory diagnostic-tests.

1.3 Methods

Of 1,000 anginal patients assessed for quality of life and 1,350 patients assessed for peripheral vascular disease with diagnostic laboratory tests items response modeling was applied using the Latent Trait Analysis program −2 of Uebersax.

1.4 Results

The 32 different response patterns obtained from test batteries of five items produced 32 different quality of life scores ranging from 3.4 to 74.5% and 32 different levels peripheral vascular disease ranging from 9.9 to 83.5% with overall mean scores, by definition, of 50%, while the classical method for analysis only produced the discrete scores 0–5. The item response models produced an adequate fit for the data as demonstrated by chi-square goodness of fit values/degrees of freedom of 0.86 and 0.64.

1.5   Conclusions

1. Quality of life assessments and diagnostic tests can be analyzed through item response modeling, and provide more sensitivity than do classical linear models.
2. Item response modeling can change largely qualitative data into fairly accurate quantitative data, and can, even with limited sets of items, produce fairly accurate frequency distribution patterns of quality of life, severity of disease, and other latent traits.

2   Introduction

Item response models are applied for analyzing item scores of psychological and intelligence tests, and they are based on exponential relationships between the psychological traits and the item responses [1, 2]. Items are usually questions with "yes" or "no" answers. Item response models were invented by Georg Rasch, a mathematician from Copenhagen who was unable to find work in his discipline in the 30ths and turned to work as a psychometrician [3]. These models are, currently, the basis for modern psychological testing including computer-assisted adaptive testing [4]. Advantages compared to classical linear testing include first that item response models do not use reliability as a measure of their applicability, but instead use formal goodness of fit tests [5]. Second, the scale does not need to be of an interval nature. As a consequence the effects of covariates can be analyzed and reported with odds ratios, independently of the item format and population averages. Ceiling effects are, therefore, much less of a problem than they are with classical linear methods [6].

Our group [7] and the group of Dr. Kessler [8] were the first to apply item response modeling to quality of life assessments. Like psychometric properties quality of life is a multidimensional construct and is often investigated in homogeneous populations. Both aspects are a direct threat to the reliability, because reliability is a direct function of the dimensionality of the item pool and of the variance of the true score in the population. Indeed, item response modeling may be suitable for quality of life analyses, although not widely used so far [9–13]. But this may be a matter of time. Quality of life (QOL) research is still in its infancy, and modern QOL batteries provide better validity and reliability [14], making it better suitable for methods like item response modeling.

Not only quality of life, but also current clinical diagnostic batteries are increasingly multidimensional, particularly, in clinical research like diagnostic test batteries in the vascular laboratory or catheterization laboratory: multiple tests are often used to assess the presence of a single disease or disease severity. To date item response modeling has not yet been applied in this field. The current chapter is the first effort for that purpose, and was also written to explain the principles of item response modeling to the readership of clinical investigators. Data examples are given of both a quality of life assessment and a diagnostic test battery in the vascular laboratory.

3   Item Response Modeling, Principles

With psychometric item response modeling the data of a test sample are exponentially modeled according to:

Probability of responding to an item (yes / no) = $e^{\text{(ability level of patient)}-\text{(difficulty level of item)}}$.

This equation can also be described as:

Log odds of responding to an item (yes / no) =
(ability level of patient) – (difficulty level of item).

Multiple items in a single test can be simply added up:

Probability of responding to a set of items (yes / no) =
$\sum e^{\text{(ability levels of patients)}-\text{(difficulty levels of items)}}$.

Log odds of responding to a set of items (yes / no) =
$\sum$ (ability levels of patients) – (difficulty levels of items).

Software is used to calculate the best fit ability parameters, otherwise called latent traits, and the best fit difficulty parameters for the data given. Then, based on these parameters, just like with logistic models for making predictions from risk factor profiles, predictions can be made about individual levels of intellectual and psychological abilities [1–6]. Similarly, predictions about levels of quality of life [7, 8] and, maybe also, severity of clinical diseases can respectively be made with quality of life data and diagnostic laboratory data.

For analysis the data are fitted within the standard Gaussian distribution. A problem is that item response modeling is not available in standard statistical software. However, for dichotomous items plenty software is commercially and freely available, Egret [15], RSP [16], OPLM [17], and Free Software LTA [18]. For polytomous items such software is rapidly being developed [19, 20]. For Windows BILOG-MG and MULTILOG are available [21]. All of the above software can handle large data files, the numbers of items to be scored are now only limited by the memory capacities of the hardware.

In the current paper, we choose to use the Free Software LTA (latent trait analysis) –2 (with binary items) of John Uebersax, 2006 [18].

The interesting things about item response modeling are

1. that they are more realistic than classical methods : e.g., with a classical model the data would produce a quality of life between 0 and 100% while patients with a quality of life of 0 and 100% in reality do not exist; in contrast, with item response modeling quality of life levels are expressed as distances from an average level;

2. that they are more flexible and precise, and, therefore, more suitable for making predictions about individual patients, for example, a set of 5 items will give 5 levels of quality of life or severity of disease in the usual classical model, with item response modeling it will give 32 levels.

The following type of data are suitable for item response modeling. A sub-domain of mental depression after a myocardial event is assessed with five items (answer yes/no): (1) not hopeful, (2) blue feeling, (3) tired in the morning, (4) worrier, (5) not talking. If we review the answers, we may observe that, for example, the items (4) and (5) are less often confirmed by our test sample subjects than the other three items. They may, therefore, be expressions of a more severe level of depression. Item response models, unlike the classical models for psychometric assessments, account for and make use of the different levels of severity of items in a test battery. By doing so they change largely qualitative data into fairly accurate quantitative data. They use for that purpose the (slight) differences between individual patients in response pattern to a set of items.

The results of the item response model are fitted to a standard normal Gaussian curve. Both the chi-square goodness of fit and the Kolmogorov-Smirnov (KS) goodness of fit test can be used to assess how closely the results actually follow the Gaussian curve, respectively using a significant chi-square value [22] and using the largest cumulative difference between observed and expected frequencies according to the KS table [23], as criteria for adequacy of the model for making predictions.

4 Quality of Life Assessment

As an example we will now analyze the five-item of a mobility-domain of a quality of life battery for patients with coronary artery disease in a group of 1,000 patients. Instead of five many more items can be included. However, for the purpose of the simplicity we will again use five items: the domain mobility in a quality of life battery was assessed by answering "yes or no" to experienced difficulty (1) while climbing stair, (2) on short distances, (3) on long distances, (4) on light household work, (5) on heavy household work. In Table 8.1 the data of 1,000 patients are summarized. These data can be fitted into a standard normal Gaussian frequency distribution curve (Fig. 8.1). From the Fig. 8.1 it can be seen that the items used here are more adequate for demonstrating low quality of life than they are for demonstrating high quality of life, but, nonetheless, an entire Gaussian distribution can be extrapolated from the data given. The lack of histogram bars on the right side of the Gaussian curve suggests that more high quality of life items in the questionnaire

Table 8.1 A summary of a 5-item mobility-domain quality of life data of 1,000 anginal patients

No. Response pattern	Response pattern (1 = yes, 2 = no) to items 1–5	Observed frequencies
1.	11111	4
2.	11112	7
3.	11121	3
4.	11122	12
5.	11211	2
6.	11212	2
7.	11221	4
8.	11222	5
9.	12111	2
10.	12112	9
11.	12121	1
12.	12122	17
13.	12211	1
14.	12212	4
15.	12221	3
16.	12222	16
17.	21111	11
18.	21112	30
19.	21121	15
20.	21122	21
21.	21211	4
22.	21212	29
23.	21221	16
24.	21222	81
25.	22111	17
26.	22112	57
27.	22121	22
28.	22122	174
29.	22211	12
30.	22212	62
31.	22221	29
32.	22222	263
		1000

would be welcome in order to improve the fit of the histogram into the Gaussian curve. Yet it is interesting to observe that, even with a limited set of items, already a fairly accurate frequency distribution pattern of all quality of life levels of the population is obtained.

The LTA-2 software program is used [18]. We enter the data file and command: Gaussian error model for IRF shape, chi-square goodness of fit for Fit Statistics, then Frequency table, and, finally, EAP score table. The software program calculates the quality of life scores of the different response patterns as EAP (expected ability a posteriori) scores. These scores can be considered as the z-values of a normal

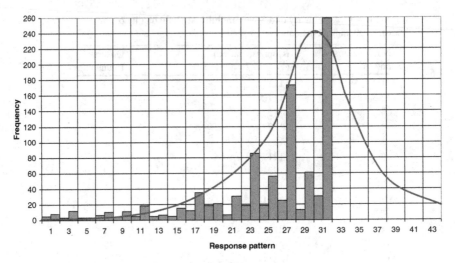

Fig. 8.1 Frequency distribution of response patterns to a five item mobility domain of a quality of life battery in 1,000 patients with coronary artery disease. The lack of histogram bars on the right side of the Gaussian curve suggests that more high quality of life items in the questionnaire would be welcome in order to improve the fit of the histogram into the Gaussian curve. Yet it is interesting to observe that even with a limited set of items already a fairly accurate frequency distribution pattern of all quality of life levels of the population is obtained

Gaussian curve, meaning that the associated areas under curve of the Gaussian curve is an estimate of the level of quality of life.

There is, approximately,

a 50% quality of life level with an EAP score of 0,
a 35% QOL level with an EAP score of −1 (standard deviations),
a 2.5%   "   "   "   of −2
a 85%   "   "   of +1
a 97.5%   "   "   of +2

In Table 8.2 the EAP scores per response pattern is given as well as the AUC (= quality of life level) values as calculated by the software program are given. In the fourth column the classical score is given ranging from 0 (no yes answers) to 5 (five yes answers).

It can be observed, that, unlike the classical scores, running from 0 to 100%, the item scores are more precise and vary from 3.4 to 74.5% with an overall mean score, by definition, of 50%. The item response model produced an adequate fit for the data as demonstrated by chi-square goodness of fit values/degrees of freedom of 0.86. What is even more important, is, that we have 32 different QOL scores instead of no more than 5 as observed with the classical score method. With 6 items the numbers of scores would even rise to 64. The interpretation is: the higher the score, the better the quality of life.

Table 8.2 Results of the item response analysis of the data from Table 8.1

No. Response pattern	Response pattern (1 = yes, 2 = no) to items 1–5	EAP scores (SDs)	AUCs (QOL levels) (%)	Classical scores (0–5)
1.	11111	−1.8315	3.4	0
2.	11112	−1.4425	7.5	1
3.	11121	−1.4153	7.8	1
4.	11122	−1.0916	15.4	2
5.	11211	−1.2578	10.4	1
6.	11212	−0.8784	18.9	2
7.	11221	−0.8600	19.4	2
8.	11222	−0.4596	32.3	3
9.	12111	−1.3872	8.2	1
10.	12112	−0.9946	16.1	2
11.	12121	−0.9740	16.6	2
12.	12122	−0.5642	28.8	3
13.	12211	−0.8377	20.1	2
14.	12212	−0.4389	33.0	3
15.	12221	−0.4247	33.4	3
16.	12222	0.0074	50.4	4
17.	21111	−1.3501	8.9	1
18.	21112	−0.9381	17.4	2
19.	21121	−0.9172	17.9	2
20.	21122	−0.4866	31.2	3
21.	21211	−0.7771	21.8	2
22.	21212	−0.3581	35.9	3
23.	21221	−0.3439	36.7	3
24.	21222	0.1120	54.4	4
25.	22111	−0.8925	18.7	2
26.	22112	−0.4641	32.3	3
27.	22121	−0.4484	32.6	3
28.	22122	0.0122	50.4	4
29.	22211	−0.3231	37.5	3
30.	22212	0.1322	55.2	4
31.	22221	0.1433	55.6	4
32.	22222	0.6568	74.5	5

EAP expected ability a posteriori, *QOL* quality of life

5 Clinical and Laboratory Diagnostic-Testing

So far item response modeling has not been used for clinical diagnostic procedures. Suppose that instead of 5 items predicting about quality of life to be answered with a yes or no, we have 5 vascular-laboratory tests predicting about the presence of peripheral vascular disease, for example (1) an ankle pressure < brachial pressure, (2) a reduction of ankle pressure after 5 min treadmill >25%, (3) a proximal thigh pressure > 35 mmHg below brachial, (4) segmental pressures (thigh, calf, ankle) > 20 mmHg difference, and (5) a toe pressure < 35% different from brachial. Similarly to the above sample, response patterns can be obtained from a data file of patients in analysis for peripheral vascular disease. The classical and item response scores from the response patterns tell us something about the expected magnitude of the vascular disease, but the item response model performs better than the classical score, because the classical score only gives the numbers of positive tests as estimators, while the item response model gives 32 levels injury. The larger the score, the more severe the disease. In 1,350 patients the predictors were measured. The data file was entered in the above LTA-2 analysis program [18]. In Table 8.3 the results of the analysis is given. The areas under the curve present the item response scores. They run from 9.9 to 83.5%. For each response pattern a separate score is produced by the analysis. The item response model produced an adequate fit for the data as demonstrated by chi-square goodness of fit values/degrees of freedom of 0.64.

When using the above item response scores and classical scores in simulated trials, it is observed, as expected, that the item response score method provides a much better sensitivity to demonstrate significant effects than does the classical score method (Table 8.4). This is so both with parallel-group and crossover designs.

6   Discussion

Do we have to assess item response models for reliability/validity? In the above laboratory test example the items were based on add-up sums of previously validated predictors. However, otherwise, they are used here in a somewhat different context, and psychological and quality of life items are generally not based on previously validated predictors. An important advantage of item response models is that they, strictly, do not require parallel tests for such purposes. This is, because the data are internally "sort of" tested for reliability:

1. ability levels of patients are tested against difficulty levels of items,
2. quality of life levels of patients are tested against quality of life levels of the items,
3. health levels of patients are tested against the health predicting levels of the items.

Yet both patient-levels and item-levels are unknown, but they have a meaningful interaction, and this interaction is mainly measured with item response modeling.

A problem is, of course, that the data should fit the Gaussian distribution, but various goodness of fit tests are available for that purpose. The software program applied in the above example makes by default use of chi-square goodness of fit tests, and the analysis does not proceed if an adequate goodness of fit is not obtained.

Logistic models for predicting risk factors, based on exponential relationships between the predictor and the outcome, have been demonstrated to provide an adequate fit for risk profiling in clinical disease and other fields, and seems to better

Table 8.3 Results of item response analysis of five laboratory predictors of peripheral vascular disease in 1,350 patients

No. response pattern	Response pattern (1 = yes, 2 = no) to items 1–5	EAP scores (SDs)	AUCs (severity of disease levels) (%)	Classical scores (0–5)
1.	11111	−1.97	12.4	0
2.	11112	−1.61	15.4	1
3.	11121	−2.04	12.1	1
4.	11122	−1.55	16.1	2
5.	11211	−1.73	14.2	1
6.	11212	−1.35	18.9	2
7.	11221	−0.85	19.1	2
8.	11222	−0.46	32.3	3
9.	12111	−1.55	16.1	1
10.	12112	−1.34	9.9	2
11.	12121	−0.80	21.1	2
12.	12122	−0.74	23.0	3
13.	12211	−0.61	27.2	2
14.	12212	−0.44	33.0	3
15.	12221	−0.51	33.4	3
16.	12222	−1.55	16.1	4
17.	21111	−0.56	28.9	1
18.	21112	−0.35	36.2	2
19.	21121	−0.31	37.9	2
20.	21122	0.00	50.2	3
21.	21211	0.00	50.1	2
22.	21212	−0.05	48.1	3
23.	21221	0.23	59.1	3
24.	21222	0.26	60.0	4
25.	22111	0.47	68.2	2
26.	22112	0.06	52.3	3
27.	22121	0.60	72.6	3
28.	22122	0.01	50.4	4
29.	22211	−0.32	37.5	3
30.	22212	0.69	75.3	4
31.	22221	0.73	76.6	4
32.	22222	0.97	83.5	5

EAP expected ability a posteriori, *AUC* area under the curve of Standard Gaussian curve, an estimate of the severity of a disease

fit data than does linear modeling. The same may be true with binary quality of life and diagnostic laboratory data. However, more studies are required. Yet, we believe that item response modeling has great potential for improving the accuracy and precision of quality of life and laboratory research.

What is even more important, it does enable to make more exact estimations in individual patients than classical methods do. Also, being more precise and sensitive, it should be more so to estimate patients that have missing data.

Table 8.4 When using the item response scores and classical scores from Table 8.3 in simulated trials, it is observed, as expected, that the item response score method provides a much better sensitivity to demonstrate significant effects than does the classical score method. This is so both with parallel-group and crossover designs

Parallel-group study

	Item response scores		Classical scores	
	Group 1	Group 2	Group 1	Group 2
	16.1	33.4	2	3
	18.9	28.9	2	1
	32.3	36.2	3	2
	9.9	50.2	2	3
	23.0	48.1	3	3
	33.0	60.0	4	4
	16.1	52.3	4	3
	36.2	50.4	2	4
	50.2	75.3	3	4
	48.1	83.5	3	5
Mean scores	28.38	51.83	2.7	3.2
Standard deviation	13.843	17.477	0.6749	1.1353
Mean difference		23.45		0.50
Standard error		7.05		0.42
		t-value$=3.1$		t-value$=1.19$
		$p<0.01$		not significant

Crossover study

Patient no	Item response scores		Differences	Classical scores		Differences
1.	16.1	33.4	−17.3	2	3	−1
2.	18.9	28.9	−10.0	2	1	1
3.	32.3	36.2	−3.9	3	2	1
4.	9.9	50.2	−40.3	2	3	−1
5.	23.0	48.1	−25.1	3	3	0
6.	33.0	60.0	−27.0	3	4	−1
7.	16.1	52.3	−36.2	4	3	1
8.	36.2	50.4	−14.2	2	4	−2
9.	50.2	75.3	−25.2	3	4	−1
10.	48.1	83.5	−35.4	3	5	−2
Mean difference			−23.46			−0.7
Standard error			3.79			0.9
			t-value$=6.19$			t-value$=0.78$
			$p<0.002$			not significant

We should discuss some limitation of the novel approach. Unlike the classical methods item response modeling is an invariant method, which means that each item is applied as point estimate without variance. This explains part of the sensitivity of the method, but at the same time means that some dispersion in the data is at risk of being underestimated. However, invariant tests are common in physics, and have received increasing attention in clinical research. E.g., Fisher exact tests and log likelihood ratio tests are of this kind. In addition, with item response modeling it may be, particularly, justified not to include variance, calculated as (squared) distances from mean scores, because the individual data are scored against a continuum of scores.

We hope that this brief introduction will stimulate clinical investigators to pay more attention to the possibilities item response modeling offers.

7 Conclusions

Item response models using exponential modeling are more sensitive than classical linear methods for making predictions from psychological questionnaires. This chapter is to assess whether they can also be used for making predictions from quality of life questionnaires and clinical and laboratory diagnostic-tests.

Of 1,000 anginal patients assessed for quality of life and 1,350 patients assessed for peripheral vascular disease with diagnostic laboratory tests items response modeling was applied using the Latent Trait Analysis program −2 of Uebersax.

The 32 different response patterns obtained from test batteries of 5 items produced 32 different quality of life scores ranging from 3.4 to 74.5% and 32 different levels peripheral vascular disease ranging from 9.9 to 83.5% with overall mean scores, by definition, of 50%, while the classical method for analysis only produced the discrete scores 0–5. The item response models produced an adequate fit for the data as demonstrated by chi-square goodness of fit values/degrees of freedom of 0.86 and 0.64.

7.1 We Conclude

3. Quality of life assessments and diagnostic tests can be analyzed through item response modeling, and provide more sensitivity than do classical linear models.
4. Item response modeling can change largely qualitative data into fairly accurate quantitative data, and can, even with limited sets of items, produce fairly accurate frequency distribution patterns of quality of life, severity of disease, and other latent traits.

References

1. Baker FB, Kim SH (2004) Item response theory: parameter estimation techniques. Marcel Dekker, New York
2. De Boeck P, Wilson M (2004) Explanatory item response models. A generalized linear and nonlinear approach. Springer, New York

3. Rasch G (1980) Probabilistic models from intelligence and attainment tests, expanded edn. The University of Chicago Press, Chicago
4. Van der Linden WJ, Veldkamp BP (2004) Constraining item exposure in computer adaptive testing with shadow tests. J Edu Behav Stat 29:273–291
5. Zwinderman AH (1991) A generalized Rasch model with manifest predictors. Psychometrika 33:AS 109–AS 119
6. Fischer GH (1974) Einfuhrung in die theorie psychologischer tests. Huber, Bern
7. Zwinderman AH, Niemeijer MG, Kleinjans HA, Cleophas TJ (1998) Application of item response modeling for quality of life assessment. In: Kuhlmann J, Mrozikiewicz A (eds) Clinical pharmacology vol 16. What should a clinical pharmacologist know to start a clinical trial (phase I and II). Zuckschwerd Verlag, Munich, pp 48–55
8. Kessler RC, Mrocek DK (1995) Measuring the effects of medical interventions. Med Care 33:109–119
9. Uttaro T, Lehman A (1999) Graded response modeling of the quality of life interview. Eval Program Plann 22:41–52
10. Douglas JA (1999) Item response models for longitudinal quality of life data in clinical trials. Stat Med 18:2917–2931
11. Reeve BB, Hays RD, Chang CH, Perfetto E (2007) Applying item response theory to enhance health outcomes. Qual Life Res 16(s1):1–3
12. Teresi JA, Fleishman JA (2007) Differential item functioning and health. Qual Life Res 16(s1):33–42
13. Cook KF, Teal CR, Bjorner JB, Celia D et al (2007) Item response theory health outcomes data analysis project: an overview and summary. Qual Life Res 16(s1):121–132
14. Cleophas TJ, Zwinderman AH, Cleophas TF, Cleophas EP (2009) Lack of sensitivity of quality of life (QOL) assessments. In: Statistics applied to clinical trials, 4th edn. Springer, Dordrecht, pp 323–329
15. Anonymous (1991) Egret manual. Statistics and epidemiology research computing. University of Seattle, Seattle
16. Glas CA, Ellis J (1993) Rasch scaling program (RSP) I.E.C. University of Groningen, Groningen
17. Verhelst ND (1993) One parameter logistic model (OPLM). CITO, Arnhem
18. Uebersax J (2006) Free software LTA (latent trait analysis) −2 (with binary items). www.john-uebersax.com/stat/Ital.htm
19. Conquest Generalized Item Response Modeling Software www.rasch.org/rmt/rmt133o.htm
20. Full Lifecycle Unified Modeling Language Modeling Software www.sparxsystems.eu/?gclid
21. Item Response Modeling with BILOG-MG and MULTILOG for Windows www.eric.ed.gov/ERICWebPortal/custom/portlets/recorDetails/detailminii.jsp
22. Cleophas TJ, Zwinderman AH, Cleophas TF, Cleophas EP (2009) Method 1: the chi -square goodness of fit test. In: Statistics applied to clinical trials, 4th edn. Springer, Dordrecht, pp 356–357
23. Cleophas TJ, Zwinderman AH, Cleophas TF, Cleophas EP (2009) Method 2: the kolmogorov-smirnov goodness of fit test. In: Statistics applied to clinical trials, 4th edn. Springer, Dordrecht, pp 357–359

Chapter 9
Time-Dependent Predictor Modeling

1 Summary

1.1 Background

Individual patients' predictors of survival may change across time, because people may change their lifestyles. Standard statistical methods do not allow adjustments for time-dependent predictors. In the past decade time-dependent factor analysis has been introduced as a novel approach adequate for the purpose.

1.2 Objective

Using examples from survival studies we assess the performance of the novel method. SPSS statistical software is used.

1.3 Methods and Results

1. Cox regression is a major simplification of real life: it assumes that the ratio of the risks of dying in parallel groups is constant over time. It is, therefore, inadequate to analyze, for example, the effect of elevated LDL cholesterol on survival, because the relative hazard of dying is different in the first, second and third decade. The time-dependent Cox regression model allowing for non-proportional hazards is applied, and provides a better precision than the usual Cox regression (p-value 0.117 versus 0.0001).
2. Elevated blood pressure produces the highest risk at the time it is highest. An overall analysis of the effect of blood pressure on survival is not significant, but,

after adjustment for the periods with highest blood pressures using the segmented time-dependent Cox regression method, blood pressures is a significant predictor of survival (p = 0.04).
3. In a long term therapeutic study treatment modality is a significant predictor of survival, but after the inclusion of the time-dependent LDL-cholesterol variable, the precision of the estimate improves from a p-value of 0.02 to 0.0001.

1.4   Conclusions

1. Predictors of survival may change across time, e.g., the effect of smoking, cholesterol, and increased blood pressure in cardiovascular research, and patients' frailty in oncology research.
2. Analytical models for survival analysis adjusting such changes are welcome.
3. The time-dependent and segmented time-dependent predictors are adequate for the purpose.
4. The usual *multiple* Cox regression model can include both time-dependent and time-independent predictors.

2   Introduction

Assessing the effects of health predictors on morbidity / mortality is an important objective in clinical research. Usually, the individual patients' values of a health predictor are evaluated at the time of entry in a study, and the final effects on morbidity / mortality are collected years later. Logistic and Cox regression models are commonly applied to determine, whether the health predictors significantly contributed to the risk of events / hazard of deaths etc [1]. The problem with this approach is the assumption, that the individual patients' values of the health predictors do not change across time. This may be true for short time observations in simple creatures like mosquitoes. However, humans are more complex and creative, and tend to change their lifestyles in the course of time. It would mean, for example, that the risk of smoking on death cannot be estimated from the numbers of cigarettes at the time of entry, if people tend to give up smoking while on trial. Therefore, an ongoing adjustment of the values of risk factors during the time of observation would be a more adequate assessment. However, standard statistical methods do not allow for such adjustments. In 1996 the group of Abrahamowicz [2] was the first to present a model for time-dependent factor analysis based on the traditional Cox regression model. It is now available in SPSS [3] statistical software and other major software programs, but, unfortunately, still rarely applied. The current chapter explains the novel model using examples from survival studies, and was written to assess the performance of the novel method, and to familiarize the clinical research community with this important approach for improved survival analysis.

Fig. 9.1 Exponential curve
of survival of mosquitoes

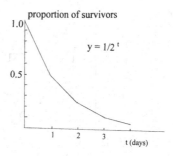

3 **Cox Regression Without Time-Dependent Predictors**

Cox regression is immensely popular. It uses exponential models: per time unit the
same percentage of patients has an event. This exponential model may be adequate
for survival of mosquitoes, they usually die whenever they collide: in a room full of
mosquitoes after the 1st day 50% may be alive, after the 2nd day 25%, etc. (Fig. 9.1).
However, human beings are much more complex and creative, and do not usually
die from collisions. Yet, this exponential model is widely applied for the comparison
of Kaplan-Meier curves in humans.

Cox regression uses an exponential model according to the following equation
(t=time):

$$\text{proportion survivors} = 1/2\, t = 2^{-t}$$

In true biology: e (=2.71828) better fits data than 2, and k is used as a constant
for the species

$$\text{proportion survivors} = e^{-kt}$$

Kaplan-Meier curves are analyzed using this exponential model. Examples of
such equations are underneath:

$$\text{proportion survivors} = e^{-kt-bx}$$

x=binary variable (only 0 or 1, 0 means treatment-1, 1 means treatment-2),
b=the regression coefficient,

if x = 0, then the equation turns into proportion survivors = e^{-kt}
if x = 1, then   "   "   "   proportion survivors = e^{-kt-b}

Figure 9.2 gives an example of two Kaplan Meier curves of groups of patients
surviving cancer after the start of either treatment-1 or -2. The continuous lines give

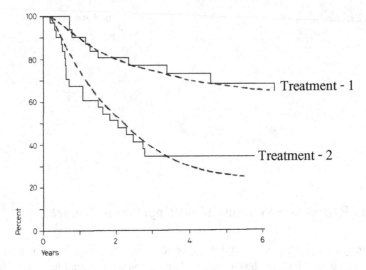

Fig. 9.2 Example of two Kaplan Meier curves of groups of patients surviving cancer after the start of either treatment-1 or -2. The *continuous lines* give the real data, the *dotted lines* the curves modeled by the Cox regression program

the real data, the dotted lines the curves modeled by the Cox regression program. We are not so much interested in precise pattern of the separate dotted curves, but rather in the ratio of the two curves (the relative chance of surviving), which is given by: $e^{-kt-b}/e^{-kt} = e^{-b}$.

Consequently, the relative risk death (the hazard ratio) is given by e^{b}.

The hazard ratio (HR) = risk death$_{treatment-2}$ / risk death$_{treatment-1}$ = e^{b}.
eLog HR (natural log) = b.

The software calculates the best fit b for the given data, if b is significantly > 0, then the HR (= antilog b) is significantly > 1, which indicates a significant difference in risk-death between treatment-2 and treatment-1. We use SPSS statistical software for analysis (var = variable).

Command: Analyze....survival....Cox regression....time: follow months.... status: var 2....define event (1)....Covariates....categorical: treatment.... Continue....plots....survival => ....hazard....continue....OK.

The analysis produces a b-value (regression coefficient) of 1.10 with a p-value of 0.01, which means that treatment modality is a significant predictor of survival: treatment-1 is much better than treatment-2 with a hazard ratio of $e^{b} = e^{1.10} = 3.00$. A nice thing with Cox regression is that like with linear and logistic regression additional x-variables can be added to the model, for example patient characteristics like age, gender, comorbidities etc.

A problem with Cox regression is that it is a major simplification of real life. It assumes that the ratio of risks of dying in the two groups is constant over time, and that this is also true for various subgroups like different age and sex groups. It is inadequate if, for example,

1. treatment effect only starts after 1–2 years,
2. treatment effect starts immediately(coronary intervention),
3. unexpected effect starts to interfere (graft-versus-host).

Why are these situations inadequate? This is, because the value of the treatment effect changes over time, and this does not happen in a fashion that is a direct function of time. Such situations thus call for an approach that adjusts for the time-dependency. Covariates other than treatment modalities, like elevated LDL cholesterol, and hypertension in cardiovascular research or patients' frailty in oncology research are similarly at risk of changing across time, and do, likewise, qualify for the underneath alternative approach. In the underneath sections the alternative analysis is explained.

4   Cox Regression with a Time-Dependent Predictor

The level of LDL cholesterol is a strong predictor of cardiovascular survival. However, in a survival study virtually no one will die from elevated values in the first decade of observation. LDL cholesterol may be, particularly, a killer in the second decade of observation. Then, in the third decade those with high levels may all have died, and other reasons for dying may occur. In other words the deleterious effect of 10 years elevated LDL-cholesterol may be different from that of 20 years. The Cox regression model is not appropriate for analyzing the effect of LDL cholesterol on survival, because it assumes that the relative hazard of dying is the same in the first, second and third decade. Thus, there seems to be a time-dependent disproportional hazard, and if you want to analyze such data, an extended Cox regression model allowing for non-proportional hazards can be applied [4], and is available in SPSS [3] statistical software. In the underneath example 60 patients are followed for 30 years for the occurrence of a cardiovascular event. Each row represents a patient, the columns are the patient characteristics, otherwise called the variables.

Variable (Var)					
1	2	3	4	5	6
1,00	1	0	65,00	0,00	2,00
1,00	1	0	66,00	0,00	2,00
2,00	1	0	73,00	0,00	2,00
2,00	1	0	54,00	0,00	2,00
2,00	1	0	46,00	0,00	2,00
2,00	1	0	37,00	0,00	2,00

(continued)

(continued)

Variable (Var)					
1	2	3	4	5	6
2,00	1	0	54,00	0,00	2,00
2,00	1	0	66,00	0,00	2,00
2,00	1	0	44,00	0,00	2,00
3,00	0	0	62,00	0,00	2,00
4,00	1	0	57,00	0,00	2,00
5,00	1	0	43,00	0,00	2,00
6,00	1	0	85,00	0,00	2,00
6,00	1	0	46,00	0,00	2,00
7,00	1	0	76,00	0,00	2,00
9,00	1	0	76,00	0,00	2,00
9,00	1	0	65,00	0,00	2,00
11,00	1	0	54,00	0,00	1,00
12,00	1	0	34,00	0,00	1,00
14,00	1	0	45,00	0,00	1,00
16,00	1	0	56,00	1,00	1,00
17,00	1	0	67,00	1,00	1,00
18,00	1	0	86,00	1,00	1,00
30,00	1	0	75,00	1,00	2,00
30,00	1	0	65,00	1,00	2,00
30,00	1	0	54,00	1,00	2,00
30,00	1	0	46,00	1,00	2,00
30,00	1	0	54,00	1,00	2,00
30,00	1	0	75,00	1,00	2,00
30,00	1	0	56,00	1,00	2,00
30,00	1	1	56,00	1,00	2,00
30,00	1	1	53,00	1,00	2,00
30,00	1	1	34,00	1,00	2,00
30,00	1	1	35,00	1,00	2,00
30,00	1	1	37,00	1,00	2,00
30,00	1	1	65,00	1,00	2,00
30,00	1	1	45,00	1,00	2,00
30,00	1	1	66,00	1,00	2,00
30,00	1	1	55,00	1,00	2,00
30,00	1	1	88,00	1,00	2,00
29,00	1	1	67,00	1,00	1,00
29,00	1	1	56,00	1,00	1,00
29,00	1	1	54,00	1,00	1,00
28,00	0	1	57,00	1,00	1,00
28,00	1	1	57,00	1,00	1,00
28,00	1	1	76,00	1,00	1,00
27,00	1	1	67,00	1,00	1,00
26,00	1	1	66,00	1,00	1,00
24,00	1	1	56,00	1,00	1,00
23,00	1	1	66,00	1,00	1,00
22,00	1	1	84,00	1,00	1,00

(continued)

(continued)

Variable (Var)					
1	2	3	4	5	6
22,00	0	1	56,00	1,00	1,00
21,00	1	1	46,00	1,00	1,00
20,00	1	1	45,00	1,00	1,00
19,00	1	1	76,00	1,00	1,00
19,00	1	1	65,00	1,00	1,00
18,00	1	1	45,00	1,00	1,00
17,00	1	1	76,00	1,00	1,00
16,00	1	1	56,00	1,00	1,00
16,00	1	1	45,00	1,00	1,00

Table 9.1 Result of usual Cox regression using the variable VAR0006 (elevated LDL-cholesterol or not) as predictor and survival as outcome

	Variables in the equation					
	B	SE	Wald	df	Sig.	Exp(B)
VAR00006	−,482	,307	2,462	1	,117	,618

Var 00001 = follow-up period (years) (Var = variable)
Var 00002 = event (1 or 0, event or lost for follow-up = censored)
Var 00003 = treatment modality (0 = treatment-1, 1 = treatment-2)
Var 00004 = age (years)
Var 00005 = gender (0 or 1, male or female)
Var 00006 = LDL-cholesterol (2 or 1, < 3.9 or >= 3.9 mmol/l)

First, a usual Cox regression is performed with LDL-cholesterol as predictor of survival (var = variable).

Command: Analyze....survival....Cox regression....time: follow months.... status: var 2....define event (1)....Covariates....categorical: elevated LDL-cholesterol (Var 00006) => categorical variables....continue....plots.... survival => ....hazard....continue....OK.

The Table 9.1 shows that elevated LDL-cholesterol is not a significant predictor of survival with a p-value as large as 0.117 and a hazard ratio of 0.618. In order to assess, whether elevated LDL-cholesterol adjusted for time has an effect on survival, a time-dependent Cox regression will be performed. For that purpose the time-dependent covariate is defined as a function of both the variable time (called " T_" in SPSS) and the LDL-cholesterol-variable, while using the product of the two. This product is applied as the "time-dependent" predictor of survival, and a usual Cox model is, subsequently, performed (Cov = covariate).

Command: Analyze....survival....Cox w/Time-Dep Cov....Compute Time-Dep Cov....Time (T_) => in box Expression for T_Cov....add the sign * ....add the LDL-cholesterol variable....model....time: follow months....status: var 00002....?: define event:1....continue....T_Cov => in box covariates....OK.

Table 9.2 Result of the time-dependent Cox regression using the variable VAR0006 (elevated LDL-cholesterol or not) as a time-dependent predictor and survival as outcome

	Variables in the equation					
	B	SE	Wald	df	Sig.	Exp(B)
T_COV_	−,131	,033	15,904	1	,000	,877

The Table 9.2 shows that elevated LDL-cholesterol after adjustment for differences in time is a highly significant predictor of survival. If we look at the actual data of the file, we will observe that, overall, the LDL-cholesterol variable is not an important factor. But, if we look at the cholesterol values per decade of follow-up separately, then it is observed that something very special is going on: in the first decade virtually no one with elevated LDL-cholesterol dies. In the second decade virtually everyone with an elevated LDL-cholesterol does: LDL cholesterol seems to be particularly a killer in the second decade. Then, in the third decade other reasons for dying seem to have occurred.

5 Cox Regression with a Segmented Time-Dependent Predictor

Some variables may have different values at different time periods. For example, elevated blood pressure may be, particularly, harmful not after decades but at the very time-point it is highest. The blood pressure is highest in the first and third decade of the study. However, in the second decade it is mostly low, because the patients were adequately treated at that time. For the analysis we have to use the socalled logical expressions. They take the value 1, if the time is true, and 0, if false. Using a series of logical expressions, we can create our time-dependent predictor, that can, then, be analyzed by the usual Cox model. In the underneath example 60 patients are followed for 30 years for the occurrence of a cardiovascular event. Each row represents again a patient, the columns are the patient characteristics.

Var 1	2	3	4	5	6	7
1,00	1	65	,00	135,00	.	.
1,00	1	66	,00	130,00	.	.
2,00	1	73	,00	132,00	.	.
2,00	1	54	,00	134,00	.	.
2,00	1	46	,00	132,00	.	.
2,00	1	37	,00	129,00	.	.
2,00	1	54	,00	130,00	.	.
2,00	1	66	,00	132,00	.	.
2,00	1	44	,00	134,00	.	.
3,00	0	62	,00	129,00	.	.
4,00	1	57	,00	130,00	.	.
5,00	1	43	,00	134,00	.	.
6,00	1	85	,00	140,00	.	.

(continued)

(continued)

Var 1	2	3	4	5	6	7
6,00	1	46	,00	143,00	.	.
7,00	1	76	,00	133,00	.	.
9,00	1	76	,00	134,00	.	.
9,00	1	65	,00	143,00	.	.
11,00	1	54	,00	134,00	110,00	.
12,00	1	34	,00	143,00	111,00	.
14,00	1	45	,00	135,00	110,00	.
16,00	1	56	1,00	123,00	103,00	.
17,00	1	67	1,00	133,00	107,00	.
18,00	1	86	1,00	134,00	108,00	.
30,00	1	75	1,00	134,00	102,00	134,00
30,00	1	65	1,00	132,00	121,00	126,00
30,00	1	54	1,00	154,00	119,00	130,00
30,00	1	46	1,00	132,00	110,00	131,00
30,00	1	54	1,00	143,00	120,00	132,00
30,00	1	75	1,00	123,00	123,00	133,00
30,00	1	56	1,00	130,00	124,00	130,00
30,00	1	56	1,00	130,00	116,00	129,00
30,00	1	53	1,00	134,00	130,00	128,00
30,00	1	34	1,00	126,00	110,00	127,00
30,00	1	35	1,00	130,00	115,00	133,00
30,00	1	37	1,00	132,00	125,00	134,00
30,00	1	65	1,00	134,00	124,00	133,00
30,00	1	45	1,00	126,00	116,00	132,00
30,00	1	66	1,00	132,00	129,00	131,00
30,00	1	55	1,00	128,00	111,00	130,00
30,00	1	88	1,00	134,00	120,00	132,00
29,00	1	67	1,00	126,00	121,00	131,00
29,00	1	56	1,00	133,00	122,00	129,00
29,00	1	54	1,00	127,00	120,00	128,00
28,00	0	57	1,00	132,00	119,00	130,00
28,00	1	57	1,00	128,00	118,00	131,00
28,00	1	76	1,00	134,00	120,00	132,00
27,00	1	67	1,00	132,00	121,00	130,00
26,00	1	66	1,00	128,00	119,00	129,00
24,00	1	56	1,00	126,00	113,00	128,00
23,00	1	66	1,00	130,00	117,00	131,00
22,00	1	84	1,00	131,00	117,00	133,00
22,00	0	56	1,00	129,00	118,00	132,00
21,00	1	46	1,00	129,00	119,00	131,00
20,00	1	45	1,00	131,00	110,00	.
19,00	1	76	1,00	130,00	111,00	.
19,00	1	65	1,00	134,00	112,00	.
18,00	1	45	1,00	126,00	113,00	.
17,00	1	76	1,00	129,00	114,00	.
16,00	1	56	1,00	131,00	106,00	.
16,00	1	45	1,00	130,00	110,00	.

Table 9.3 Result of Cox regression with a segmented time-dependent predictor constructed with the relevant blood pressures from three decades of the study and survival as outcome

			Variables in the equation			
	B	SE	Wald	df	Sig.	Exp(B)
T_COV_	−,066	,032	4,238	1	,040	,936

Var 00001 = follow-up period years (Var = variable)
Var 00002 = event (0 or 1, event or lost for follow-up = censored)
Var 00003 = age (years)
Var 00004 = gender
Var 00005 = mean blood pressure in the first decade
Var 00006 = mean blood pressure in the second decade
Var 00007 = mean blood pressure in the third decade

The above data file shows that in the second and third decade an increasing number of patients have been lost. The following time-dependent covariate has been constructed for the analysis of these data (* = sign of multiplication):

$$(T\_ >= 1 \& T\_ < 11) * Var\ 5 + (T\_ >= 11 \& T\_ < 21)$$
$$*Var\ 6 + (T\_ >= 21 \& T\_ < 31) * Var\ 7$$

This predictor is entered in the usual way with the commands (Cov = covariate):

Model....time: follow months....status: var 00002....?: define event: 1 – continue....T_Cov => in box covariates....OK.

The Table 9.3 shows that, indeed, a mean blood pressure after adjustment for difference in decades is a significant predictor of survival at p = 0.040, and with a hazard ratio of 0.936 per mmHg. In spite of the better blood pressures in the second decade, blood pressure is a significant killer in the overall analysis.

6 Multiple Cox Regression with a Time-Dependent Predictor

Time-dependent predictors can be included in multiple Cox regression analyses together with time-independent predictors. The example of Sect. 2 is used once more. The data of the effects of two treatments on mortality / morbidity are evaluated using a Cox regression model with treatment modality, Var 00003, as predictor. Table 9.4 shows that treatment modality is a significant predictor of survival: the patients with treatment-2 live significantly shorter than those with treatment-1 with a hazard ratio of 0.524 and a p-value of 0.017. Based on the analysis in the above Sect. 3, we can not exclude that this result is confounded with the time-dependent predictor LDL-cholesterol. For the assessment of this question a multiple Cox model is used with both treatment modality (Var 00003) and the time-dependent LDL-cholesterol (T_Cov) as predictors of survival.

Table 9.4 Simple Cox regression with treatment modality as predictor and survival as outcome

	\multicolumn{6}{c}{Variables in the equation}					
	B	SE	Wald	df	Sig.	Exp(B)
VAR00003	−,645	,270	5,713	1	,017	,524

Table 9.5 Multiple Cox regression with treatment modalities and the time-dependent LDL-cholesterol predictor as predictor variables and survival as outcome

	\multicolumn{6}{c}{Variables in the equation}					
	B	SE	Wald	df	Sig.	Exp(B)
VAR00003	−1,488	,365	16,647	1	,000	,226
T_COV_	−,092	,017	29,017	1	,000	,912

Table 9.5 shows that both variables are highly significant predictors independent of one another. A hazard ratio of only 0.226 of one treatment versus the other is observed. This result is not only more spectacular but also more precise given the better p-value, than that of the unadjusted assessment of the effect of treatment modality. Obviously, the time-dependent predictor was a major confounder which has now been adjusted.

7 Discussion

Transient frailty and changes in lifestyle may be time-dependent predictors in log- term research. Some readers may find it hard to understand how to code a time-dependent predictor for Cox regression. Particularly in the SPSS program the term T_ is not always well understood. T_ is actually the current time, which is only relevant for a case that is of this time. There are, generally, two types of time-dependent predictors that investigators want to use. One involves multiplying a predictor by time in order to test the proportional hazard function or fit a model with non-proportional hazards. Compute the time-dependent predictor as: T_*covariate (*=sign of multiplication). This produces a new variable which is analyzed like with the usual Cox proportional hazard method.

The other kind of time-dependent predictor is called the segmented time-dependent predictor. It is a predictor where the value may change over time, but not in a fashion that is a direct function of time. For example, in the data file from Sect. 4 the blood pressures change over time, going up and down in a way that is not a consistent function of the time. The new predictor is set up so that for each possible time interval one of the three period variables will operate.

Currently, clinical investigators increasingly perform their own data-analysis without the help of a professional statistician. User-friendly software like SPSS is available for the purpose. Also, more advanced statistical methods are possible. The time-dependent Cox regression is more complicated than the fixed time-independent

Cox regression. However, as demonstrated above, it can be readily performed by non-mathematicians along the procedures as described above.

We recommend that researchers, particularly those, whose results do not confirm their prior expectations, perform more often the extended Cox regression models, as explained in this chapter. We conclude.

1. Many predictors of survival change across time, e.g., the effect of smoking, cholesterol, and increased blood pressure in cardiovascular research, and patients' frailty in oncology research.
2. Analytical models for survival analysis adjusting such changes are welcome.
3. The time-dependent and segmented time-dependent predictors are adequate for the purpose.
4. The usual multiple Cox regression model can include both time-dependent and time-independent predictors.

8 Conclusions

Individual patients' predictors of survival may change across time, because people may change their lifestyles. Standard statistical methods do not allow adjustments for time-dependent predictors. In the past decade time-dependent factor analysis has been introduced as a novel approach adequate for the purpose.

Using examples from survival studies we assess the performance of the novel method. SPSS statistical software is used.

1. Cox regression is a major simplification of real life: it assumes that the ratio of the risks of dying in parallel groups is constant over time. It is, therefore, inadequate to analyze, for example, the effect of elevated LDL cholesterol on survival, because the relative hazard of dying is different in the first, second and third decade. The time-dependent Cox regression model allowing for non-proportional hazards is applied, and provides a better precision than the usual Cox regression (p-value 0.117 versus 0.0001).
2. Elevated blood pressure produces the highest risk at the time it is highest. An overall analysis of the effect of blood pressure on survival is not significant, but, after adjustment for the periods with highest blood pressures using the segmented time-dependent Cox regression method, blood pressures is a significant predictor of survival (p = 0.04).
3. In a long term therapeutic study treatment modality is a significant predictor of survival, but after the inclusion of the time-dependent LDL-cholesterol variable, the precision of the estimate improves from a p-value of 0.02 to 0.0001.

8.1   We Conclude

1. Predictors of survival may change across time, e.g., the effect of smoking, cholesterol, and increased blood pressure in cardiovascular research, and patients' frailty in oncology research.
2. Analytical models for survival analysis adjusting such changes are welcome.
3. The time-dependent and segmented time-dependent predictors are adequate for the purpose.
4. The usual *multiple* Cox regression model can include both time-dependent and time-independent predictors.

References

1. Cox DR (1972) Regression models and life tables. J R Stat Soc 34:187–220
2. Abrahamowicz M, Mackenzie T, Esdaille JM (1996) Time-dependent hazard ratio modeling and hypothesis testing with application in lupus nephritis. J Am Stat Assoc 91:1432–1439
3. SPSS statistical software. www.SPSS.com. Accessed 18 Dec 2012
4. Fisher LD, Lin DY (1999) Time-dependent covariates in the Cox proportional-hazards regression model. Annu Rev Public Health 20:145–157

Chapter 10
Seasonality Assessments

1 Summary

1.1 Background

Seasonal patterns are assumed in many fields of medicine. However, biological processes are full of variations and the possibility of chance findings can often not be ruled out.

1.2 Objective and Methods

Using simulated data we assess whether autocorrelation is helpful to minimize chance findings and test to support the presence of seasonality.

1.3 Results

Autocorrelation required to cut time curves into pieces. These pieces were compared with one another using linear regression analysis. Four examples with imperfect data are given. In spite of substantial differences in the data between the first and second year of observation, and in spite of, otherwise, inconsistent patterns, significant positive autocorrelations were constantly demonstrated with correlation coefficients around 0.40 (SE 0.14).

T.J. Cleophas and A.H. Zwinderman, *Machine Learning in Medicine*,
DOI 10.1007/978-94-007-5824-7_10, © Springer Science+Business Media Dordrecht 2013

1.4   Conclusions

Our data suggest that autocorrelation is helpful to support the presence of seasonality of disease, and that it does so even with imperfect data.

2   Introduction

Seasonal patterns are assumed in many fields of medicine. Examples include the incidence of nosocomial infection in a hospital, seasonal variations in hospital admissions, the course of any disease through time etc. Seasonality is defined as a yearly repetitive pattern of severity or incidence of disease. Seasonality assessments are relevant, since they enable to optimize prevention and therapy. Usually, the mean differences between the data of different seasons or months are used. E.g., the number of hospital admissions in the month of January may be roughly twice that of July. However, biological processes are full of variations and the possibility of chance findings can not be fully ruled out. For a proper assessment, at least information of a second year of observation is needed, as well as information not only of the months of January and July, but also of adjacent months. All of this information included in a single statistical test would be ideal.

In the early 1960s the Cambridge UK professor of statistics Udney Yule, a former student of the famous inventor of linear regression Karl Pearson, proposed a method for that purpose, and called it autocorrelation [1]. It is a technique that cuts time curves into pieces. Pieces are compared with one another. If they significantly fit, then seasonality is suspected to be present. Autocorrelation is currently used in many fields of science including geography, ecology, sociology, econometrics, and environmental studies [2, 3]. However, in clinical research it is little used. Of 12 seasonality papers in Medline [4–15], only two [12, 15] applied autocorrelation, while the rest just reported the presence of a significant difference between the mean data of different seasons. Autocorrelation does not demonstrate significant differences between seasons, but rather significantly repetitive patterns, meaning patterns that are stronger repetitive than could happen by chance.

In the present chapter using simulated data, we assess whether autocorrelation can support seasonality of disease, and whether it works even with imperfect data. We do hope that this chapter will stimulate clinical investigators to start using this method that is helpful to support the presence of seasonality in longitudinal data.

3   Autocorrelations

Autocorrelations is a technique that cuts time curves into pieces. These pieces are, subsequently, compared with the original datacurve using linear regression analysis. In an outpatient clinic C-reactive protein values may be higher in winter than

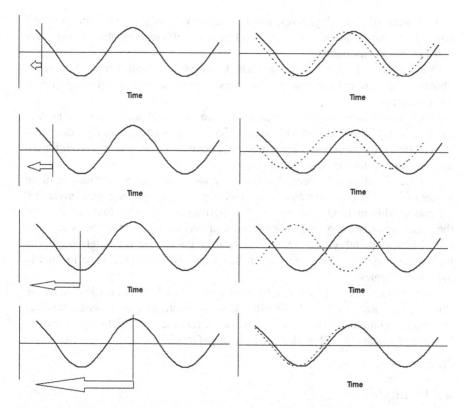

Fig. 10.1 Seasonal pattern in C-reactive protein levels in a healthy subject. Lagcurves (*dotted*) are partial copies of the datacurve moved to the left as indicated by the arrows. *First-row graphs*: the datacurve and the lagcurve has largely simultaneous positive and negative departures from the mean, and, thus, has a strong positive correlation with one another (correlation coefficient ≈ +0.6). *Second-row graphs*: this lagcurve has little correlation with the datacurve anymore (correlation coefficient ≈ 0.0). *Third-row graphs*: this lagcurve has a strong negative correlation with the data-curve (correlation coefficient ≈ −1.0). *Fourth-row graphs*: this lagcurve has a strong positive correlation with the datacurve (correlation coefficient ≈ +1.0)

in summer, and Fig. 10.1 gives a simulated example of the pattern of C-reactive protein values in a healthy subject. The curve is cut into pieces four times, and the cut pieces are called lagcurves, and they are moved to the left end of the original curve. The first-row lagcurve in Fig. 10.1 is very close to the original datacurve. When performing a linear regression analysis with the original data on the y-axis and the lagdata on the x-axis, a strong positive correlation will be found. The second lagcurve is not close anymore, and linear regression of the two curves produces a correlation coefficient of approximately zero. Then, the third lagcurve gives a mirror image of the original datacurve, and, thus, has a strong negative correlation. Finally the fourth lagcurve is in phase with the original datacurve, and, thus, has a strong positive correlation.

If, instead of a few lagcurves, monthly lagcurves are produced, then we will observe that the magnitude of the autocorrelation coefficients changes sinusoidally in the event of seasonality.

Autocorrelation coefficients significantly larger or smaller than 0 must be observed in order to conclude the presence of a statistically significant autocorrelation.

Also, *partial* autocorrelation coefficients can be calculated by comparing lag-curves of every second month with one another. This procedure should produce zero correlations subsequently to strong positive autocorrelations, and is helpful for the interpretation of the autocorrelation results.

Instead of individual values also summary measures like proportions or mean values of larger populations can be assessed for periodicity using autocorrelation. Of course, this method does not support periodicity in individual members of the populations, but it supports the presence of periodicity in populations *at large*. E.g., 24 mean monthly CRP values of a healthy population is enough to tell you with some confidence something about the spread and the chance of periodicity in these mean values.

Autocorrelations and partial autocorrelation are in SPSS Statistical Software (the module Forecasting) [16]. We will use this software program to assess whether the autocorrelation function can demonstrate a statistically significant autocorrelation, and whether it will work even with imperfect data.

4 Examples

Four simulated examples are given. Figure 10.2 is supposed to present the seasonal patterns of mean monthly C-reactive protein (CRP) values. The upper graph shows that the mean monthly CRP values in a healthy population look inconsistent. In contrast, the middle graph suggests that the pattern of bimonthly means is rather seasonal. Autocorrelation is used to assess these data. SPSS Statistical Software is used.

> We command: Analyze....Forecasting....Autocorrelations....move mean CRP into variable box....mark Autocorrelations....mark Partial Autocorrelations.... OK.

The Table 10.1 shows the autocorrelation coefficients and their standard errors. Instead of the *t*-test, the Ljung-Box test is used for statistical testing. This method assesses overall randomness on numbers of lags instead of randomness at each lag and is calculated according to

$$\text{Chi-square} = n(n+2) \, \Sigma \left[r_k^2 \, / (n-k) \right] (\text{with h degrees of freedom})$$

with n = sample size, k = lag number, Σ = add-up sum of terms between square brackets from k = 1 to k = h, with h = total number of lags tested).

Fig. 10.2 *Upper graph*: the
mean monthly CRP values
(mg/l) in a healthy population
look inconsistent. *Middle
graph*: the pattern of
bimonthly averages looks
seasonal. *Lower graph*: a
sinusoidal time-pattern of the
autocorrelation coefficients
(ACFs) is observed, with a
significant positive
autocorrelation at the months
13 with a correlation
coefficient of 0.34 (SE 0.13)

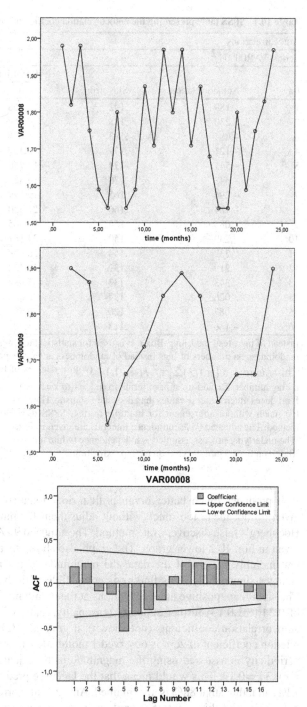

Table 10.1 SPSS table presenting the autocorrelation coefficients and their standard errors

Autocorrelations

Series: VAR00008

Lag	Autocorrelation	Std. error[a]	Box-Ljung statistic Value	df	Sig.[b]
1	,189	,192	,968	1	,325
2	,230	,188	2,466	2	,291
3	−,084	,183	2,678	3	,444
4	−,121	,179	3,137	4	,535
5	−,541	,174	12,736	5	,026
6	−,343	,170	16,825	6	,010
7	−,294	,165	19,999	7	,006
8	−,184	,160	21,320	8	,006
9	,080	,155	21,585	9	,010
10	,239	,150	24,141	10	,007
11	,237	,144	26,838	11	,005
12	,219	,139	29,322	12	,004
13	,343	,133	35,982	13	,001
14	,026	,127	36,024	14	,001
15	−,087	,120	36,552	15	0,001
16	−,166	,113	38,707	16	0,001

Instead of the t-test, the Ljung-Box test is used for statistical testing. This method assesses overall randomness on numbers of lags instead of randomness at each lag and is calculated according to Chi-square $= n\left(n+2\right)\Sigma\left[r_k^2/\left(n-k\right)\right]$ (with h degrees of freedom) with n = sample size, k = lag number, Σ = add-up sum of terms from k = 1 to k = h, with h = total number of lags tested). It produces much better p-values than does the t-statistic. However, the p-values given do not mean too much without adjustment for multiple testing. SPSS uses Hochberg's false discovery rate method. The adjusted 95% confidence intervals are given in Fig. 10.2 lower graph

[a]The underlying process assumed is independence (white noise)

[b]Based on the asymptotic chi-square approximation

It produces much better p-values than does t-statistic. However, the p-values given do not mean too much without adjustment for multiple testing. SPSS uses Hochberg's false discovery rate method. The adjusted 95% confidence intervals are given in Fig. 10.2 lower graph. The graph also shows that, in spite of the inconsistent monthly pattern of the data, the magnitude of the monthly autocorrelations changes sinusoidally. This finding is compatible with the presence of periodicity. The significant positive autocorrelations at the month no.13 (correlation coefficient of 0.34 (SE 0.13) further supports seasonality, and so does the pattern of partial autocorrelation coefficients (not shown): it gradually falls, and a partial autocorrelation coefficient of zero is observed 1 month after month 13. The strength of the periodicity is assessed using the magnitude of the squared correlation coefficient $r^2 = 0.34^2 = 0.12$. This would mean that the lagcurve predicts the datacurve by only 12%, and, thus, that 88% is unexplained. And so, autocorrelation may be significant but, in spite of this, a lot of unexplained variability, otherwise called noise, is in the data.

The Figs. 10.3, 10.4, and 10.5 give additional examples of data with imperfect time plots. Inconsistent patterns are observed in summer (Fig. 10.3) or in winter or both (Fig. 10.5), and substantial differences between the first and second winter are observed (Fig. 10.4). Similarly to the autocorrelation analysis of Fig. 10.2, all of these data give evidence for periodicity with harmonic sinusoidal patterns and with the months 10–13 well beyond the 95% confidence limit. Obviously, despite imperfect data, autocorrelation enabled to constantly support the presence of seasonality.

5 Discussion

In clinical papers seasonality has often been defined as a significant difference between the mean data of different seasons [4–11, 13, 14]. However, differences due to chance rather than true seasonal effects is possible, and the presence of seasonality is supported by testing the hypothesis of a repetitive time pattern [2, 3]. Autocorrelation is capable of doing so, and can even be successful in case of weak seasonal effects, and in the presence of a lot of noise as demonstrated in the above examples.

Autocorrelation may help suspect a seasonal disease pattern, but does not explain the cause of seasonality. E.g., affective disorders in winter [6] may partly be induced by reduced sunlight, and increased mortality rates in winter [4]. In order to confirm these assumptions, adequate information about a repetitive seasonal pattern of possible causal factors is required. If this information is available, then two analyses are possible:

1. a direct comparison of the seasonal disease patterns with the causal factor patterns using cross-correlations, a method largely similar to autocorrelations, with lagcurves of the causal factor [17];
2. multiple autocorrelation analysis with the causal factor as covariate.

(1) is available in SPSS, but few studies to date have used it. No software for (2) is available. For explaining cross-correlations, the data from Fig. 10.2 are used once more. A second variable, is added, the monthly means of the highest daily minimum temperatures (Fig. 10.6).

We command: Analyze....Forecasting....Cross-correlations....move mean monthly C-reactive values into variable box....move monthly means of highest daily minimum temperature into variable box....mark Cross-Correlations....OK.

Figure 10.6 lower graph shows that the cross-correlation is strong negative in the 13th month of observation. The cross-correlation coefficients of seven lagcurves departing from zero in both directions display a sinusoidal pattern with very significant cross-correlations at −5, 0, and +6 months. This tells us that the periodical patterns of the C-reactive proteins and those of the temperatures are pretty close, and would support a causal relationship.

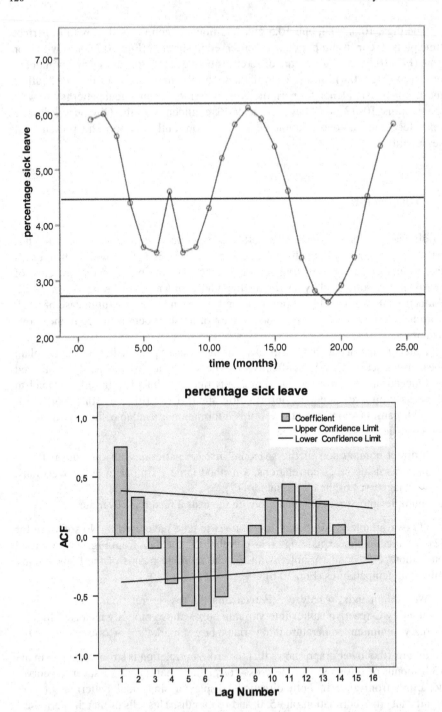

Fig. 10.3 Example of data with a considerable difference between the pattern of the first and second summer. The *lower graph* shows that, in spite of this, a sinusoidal time-pattern of the auto-correlation coefficients (ACFs) is observed, with significant positive autocorrelations at the 11 and 12 months (correlation coefficients of 0.43 (SE 0.14) and 0.42 (SE 0.14) respectively)

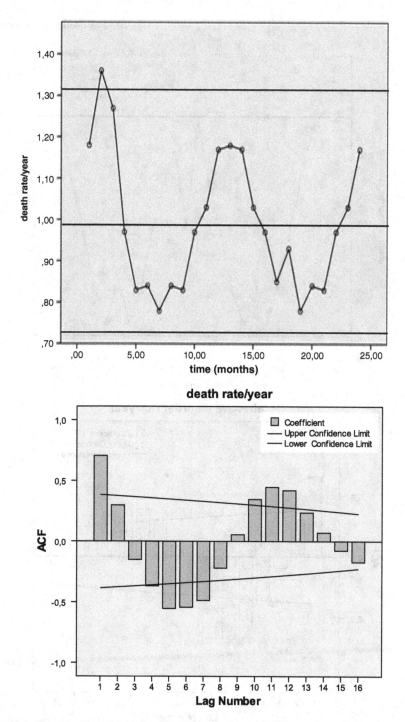

Fig. 10.4 Example where data look seasonal, but a substantial difference between the pattern in the first and that in the second winter is observed. The *lower graph* shows that, in spite of this, a sinusoidal time-pattern of the autocorrelation coefficients (ACFs) is observed, with significant positive autocorrelations at the months 10–12 with correlation coefficients of 0.35 (SE 0.15), 0.45 (SE 0.14), and 0.42 (0.14) respectively

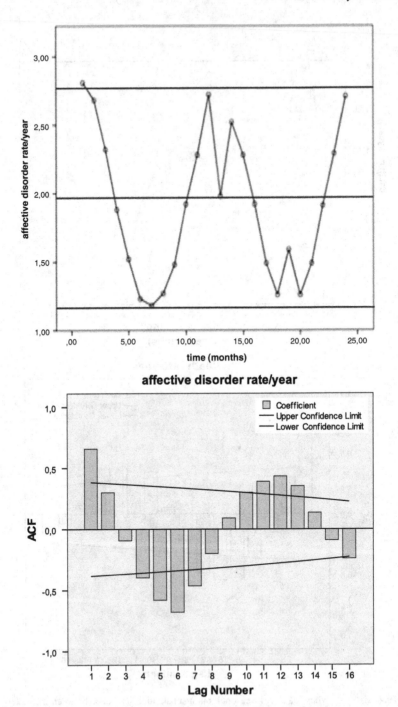

Fig. 10.5 Example where data look seasonal, but inconsistent patterns in the second winter and summer are observed. The *lower graph* shows that, in spite of this, a sinusoidal time-pattern of the autocorrelation coefficients (ACFs) is observed, with significant positive autocorrelations at the months 11–13 with correlation coefficients of 0.39 (SE 0.14), 0.43 (SE 0.14), and 0.35 (0.13) respectively

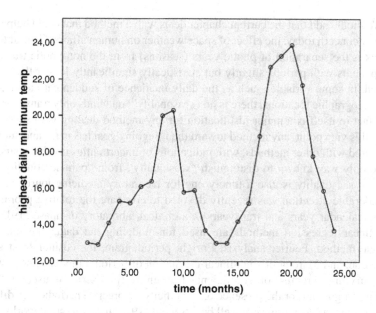

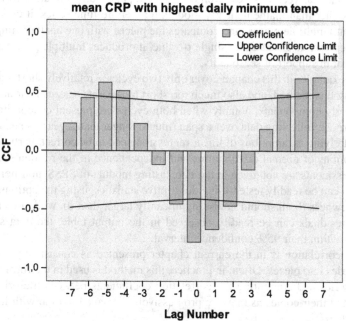

Fig. 10.6 The data from Fig. 10.2 are used once more. To the monthly mean C-reactive protein values, the monthly means of the daily minimum temperatures are added as a second variable (*upper graph*). The *lower graph* shows that the cross-correlation coefficient (CCF) is strong negative in the 13th month of observation. The cross-correlation coefficients of seven lagcurves departing from zero in both directions display a sinusoidal pattern with very significant cross-correlations at −5, 0, and +6 months. This tells that the seasonal patterns of the C-reactive proteins and those of the temperatures are pretty close, and it would support a causal relationship

We should add that the current chapter deals with a method that could help clarify a major concern today, the effect of space weather on human affairs. Central to such effects is the separation of photic years (seasons) from the non-photic transyears, components with periods slightly but statistically significantly longer than 1 year. Indeed, in some variables such as the daily incidence of sudden cardiac deaths in some geographic locations, there is no "seasonality"; one finds only transyears [18]. This fact in itself is a major justification for any method dealing with seasonality. From this viewpoint, any method toward the foregoing goal has merit and should be compared with other methods, with indications of uncertainties involved, since this is the only way known to distinguish "seasonality" from "para-seasonality". The topic of seasonality is also a timely one for laboratory medicine journals, where considerable attention was recently devoted to compare the relative prominence of the calendar years and transyears in a clinical laboratory database [19]. Often curvilinear regression methods are used for modeling the data, like the Lomb Scargle method, Fourier analysis, Enright periodogram, the cosinor least square technique, and Marquardt's nonlinear estimation of periodic models, and a significant curvilinear regression coefficient of an entire time series is used as "all or nothing" argument for the presence of periodicity. In practice periodicities different from a calendar year may very well be observed [19]. Autocorrelation works differently. Rather than mathematically modeling an entire time series, it cuts the time series into multiple pieces, and compares the pieces with one another simply using linear regression. Instead of a single p-value, it produces multiple p-values that are or are not in support of seasonality.

The examples in this chapter cover only two cycles, a relatively short duration for applying this method, and also much too short to separate seasonal variations from transyearly components, notably when both cycles are present concomitantly. For that purpose collecting data over a span (much) longer than 2 years is needed. Also, the underlying assumptions of linear regression need to be correct, particularly, the assumptions of normal distributions and independence of the residuals. Although the assessments are not given in the Forecasting module of SPSS, non-normality of the data can be readily tested using descriptive statistics, using the options Kurtosis and Skewness (both should not be significantly larger than 0), while independence of the residuals can be readily observed in the output table: the data should be largely within their 95% confidence interval.

Autocorrelation is in the current chapter presented as consistently cutting the time series into pieces. Often, in practice, this method is used in the form of getting times series and computing the correlation coefficients between the original data series and their copies as they are progressively displayed in time with longer and longer lags.

We have to mention some more limitations of the autocorrelation methodology. First, with autocorrelations linear regression is used to assess the seasonal reproducibility of disease. However, linear regression is not a strong predictor of reproducibility. E.g., the above correlation coefficients of around 0.40, although statistically significant, produce squared correlation coefficients of around $0.40^2 = 0.16$ (16%). This would mean that the lagcurve predicts the datacurve by only 16%, and, thus, that 84% is unexplained. And so, the periodicity is pretty weak, and a lot of unexplained variability, otherwise called noise, or noise due to mechanisms other than periodical reproducibility is in the data. Also, the original datacurves and the lagcurves may contain many peaks, and autocorrelation is simply based on picking the largest peaks in your curves. If these peaks happen to be bigger than the peaks due to periodicity, then this simple procedure of picking the largest peak to be the period-peak will fail. Other limitations include regression to the mean phenomena, calculations based on approximations, and the assumption of stationary time series.

The current chapter shows that autocorrelation is helpful to support the presence of seasonality of disease, and that it is so even with imperfect data. The determination of a weak periodicity calls for confirmation, but, if confirmed, it may, eventually, have relevant health consequences regarding disease prevention and treatment policies. We do hope that this chapter will stimulate clinical investigators to start using this method.

6 Conclusions

Seasonal patterns are assumed in many fields of medicine. However, biological processes are full of variations and the possibility of chance findings can often not be ruled out. Using simulated data we assess whether autocorrelation is helpful to minimize chance findings and test to support the presence of seasonality. Autocorrelation required to cut time curves into pieces. These pieces were compared with one another using linear regression analysis. Four examples with imperfect data are given. In spite of substantial differences in the data between the first and second year of observation, and in spite of, otherwise, inconsistent patterns, significant positive autocorrelations were constantly demonstrated with correlation coefficients around 0.40 (standard error 0.14). Our data suggest that autocorrelation is helpful to support the presence of seasonality of disease, and that it does so even with imperfect data.

References

1. Yates F (1952) George Udney Yule. Obit Not R Soc Stat 8:308
2. Chatfield C (2004) The analysis of time series, an introduction, 6th edn. Chapman & Hall, New York
3. Getis A (2007) Reflections on autocorrelations. Reg Sci Urban Econ J 37:491–496
4. Douglas AS, Rawles JM, Al-Sayer H, Allan TM (1991) Seasonality of disease in Kuwait. Lancet 337:1393–1397
5. Hancox JG, Sheridan SC, Feldman SR, Fleischer AB (2004) Seasonal variation of dermatologic disease in the USA: a study of office visits from 1990 to 1998. Int J Dermatol 43:8–11
6. Whitehead BS (2004) Winter seasonal affective disorder: a global biocultural perspective. ANT (Actor-Network Theory), 570. Accessed 18 Dec 2012

7. Gonzalez DA, Victora CG, Goncalves H (2008) The effects of season at time of birth on asthma and pneumonia and adulthood in a birth cohort in southern Brazil. Cad Saude Publica 24:1089–1102
8. Parslow RA, Jorm AF, Butterworth P, Jacomb PA, Rodgers B (2004) An examination of seasonality experienced by Australians living in a continental temperate climate zone. J Affect Disord 80:181–190
9. Kelly GS (2005) Seasonal variations of selected cardiovascular risk factors. Altern Med Rev 10:307–320
10. Sung KC (2006) Seasonal variation of C-reactive protein in apparently healthy Koreans. Int J Cardiol 107:338–342
11. Rudnick AR, Rumley A, Lowe GD, Strachan DP (2007) Diurnal seasonal and blood processing patterns in levels of circulating fibrinogen, fibrin D-dimer, C-reactive protein, tissue plasminogen activator, and von Willebrand factor in a 45 year old population. Circulation 115:996–1003
12. Fishman DN (2007) Seasonality of infectious diseases. Annu Rev Public Health 28:127–143
13. Lofgren E, Fefferman NH, Naumov YN, Gorski J, Naumova EN (2007) Influenza seasonality, underlying causes and modeling theories. J Virol 8:5429–5436
14. Chiriboga DE, Ma Y, Li W, Stanek EJ, Hebert JR, Merriam PA, Rawson ES, Ockene IS (2009) Seasonal and sex variation of high-sensitivity C-reactive protein in healthy adults; a longitudinal study. Clin Chem 55:313–321
15. Mianowski B, Fendler W, Szadkowska A, Baranowska A, Grzelak E, Sadon J, HKeenan H, Mlynarski W (2011) HbA1c levels in schoolchildren with type 1 diabetes are seasonally variable and dependent on weather conditions. Diabetologia 54:749–756
16. SPSS Statistical Software (2012) www.spss.com. Accessed 28 Feb 2012
17. Cleophas TJ, Zwinderman AH (2012) Time series. In: Statistics applied to clinical studies, 5th edn. Springer, New York, pp 687–693
18. Halberg F, Corneleissen G, Otsuka K, Fiser B, Mitsutake G, Wendt HW, Johnson P, Gigolashvili M, Breus T, Sonkowsky R, Chibisov SM, Katinas G, Diegelova J, Dusek J, Singh RB, Berri BL, Schwartzkopff O (2005) Incidence of sudden cardiac death, myocardial infarction and far- and near-transyears. Biomed Pharmacother 59(Suppl 1):S239–S261
19. De Andrade D, Hirata RDC, Sandrini F, Largura A, Hirata MH Uric acid biorhythm, a feature of longterm variation in a clinical database. Clin Chem Lab Med. doi:10.1515/cclm-2011-0150. Published Online 02/03/2012

Chapter 11
Non-linear Modeling

1 Summary

1.1 Background

Novel models for the assessment of non-linear data are being developed for the benefit of making better predictions from the data.

1.2 Objective

To review traditional and modern models.

1.3 Results and Conclusions

1. Logit and probit transformations are often successfully used to mimic a linear model. Logistic regression, Cox regression, Poisson regression, and Markow modeling are examples of logit transformation.
2. Either the x- or y-axis or both of them can be logarithmically transformed. Also Box Cox transformation equations and ACE (alternating conditional expectations) or AVAS (additive and variance stabilization for regression) packages are simple empirical methods often successful for linearly remodeling of non-linear data.
3. Data that are sinusoidal, can, generally, be successfully modeled using polynomial regression or Fourier analysis.
4. For exponential patterns like plasma concentration time relationships exponential modeling with or without Laplace transformations is a possibility.
5. Spline and Loess are computationally intensive modern methods, suitable for smoothing data patterns, if the data plot leaves you with no idea of the relationship

T.J. Cleophas and A.H. Zwinderman, *Machine Learning in Medicine*,
DOI 10.1007/978-94-007-5824-7_11, © Springer Science+Business Media Dordrecht 2013

between the y- and x-values. There are no statistical tests to assess the goodness of fit of these methods, but it is always better than that of traditional models.

2   Introduction

Non-linear relationships like the smooth shapes of airplanes, boats, and motor cars were constructed from scale models using stretched thin wooden strips, otherwise called splines, producing smooth curves, assuming a minimum of strain in the materials used. With the advent of the computer it became possible to replace it with statistical modeling for the purpose: already in 1964 it was introduced by Boeing [1] and General Motors [2]. Mechanical spline methods were replaced with their mathematical counterparts. A computer program was used to calculate the best fit line/curve, which is the line/curve with the shortest distance to the data. More complex models were required, and they were often laborious so that even modern computers had difficulty to process them. Software packages make use of iterations: five or more regression curves are estimated ("guesstimated"), and the one with the best fit is chosen. With large data samples the calculation time can be hours or days, and modern software will automatically proceed to use Monte Carlo calculations [3] in order to reduce the calculation times. Nowadays, many non-linear data patterns can be developed mathematically, and this chapter reviews some of them.

3   Testing for Linearity

A first step with any data analysis is to assess the data pattern from a scatter plot (Fig. 11.1).

A considerable scatter is common, and it may be difficult to find the best fit model. Prior knowledge about patterns to be expected is helpful. Sometimes, a better fit of the data is obtained by drawing y versus x instead of the reverse. Residuals of y versus x with or without adjustments for other x-values are helpful for finding a recognizable data pattern. Statistically, we test for linearity by adding a non-linear term of x to the model, particularly, x squared or square root x, etc. If the squared correlation coefficient r^2 becomes larger by this action, then the pattern is, obviously, not-linear. Statistical software like the curvilinear regression option in SPSS [4] helps you identify the best fit model. Figure 11.2 and Table 11.1 give an example. The best fit models for the data given in the Fig. 11.2 are the quadratic and cubic models.

In the next few sections various commonly used mathematical models are reviewed. The mathematical equations of these models are summarized in the appendix. They are helpful to make you understand the assumed nature of the relationships between the dependent and independent variables of the models used.

Fig. 11.1 Examples
of non-linear data sets:
(**a**) relationship between age
and systolic blood pressure,
(**b**) effects of mental stress on
fore arm vascular resistance,
(**c**) relationship between time
after polychlorobiphenyl
(PCB) exposure and PCB
concentrations in lake fish

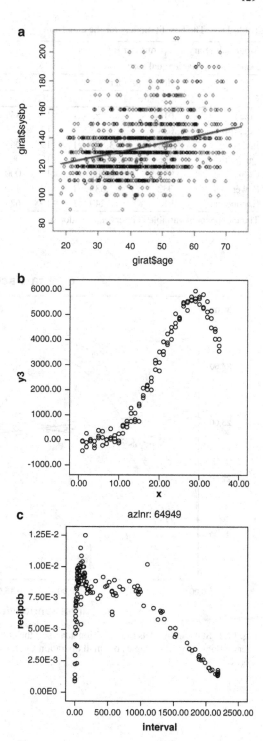

Table 11.1 The best fit models for the data from Fig. 11.2 are the quadratic and cubic models

Model summary and parameter estimates

Dependent Variable: qual care score

	Model summary					Parameter estimates			
Equation	R square	F	df1	df2	Sig.	Constant	b1	b2	b3
Linear	,018	,353	1	19	,559	25,588	−,069		
Logarithmic	,024	,468	1	19	,502	23,086	,726		
Inverse	,168	3,829	1	19	,065	26,229	−11,448		
Quadratic	,866	58,321	2	18	,000	16,259	2,017	−,087	
Cubic	,977	236,005	3	17	,000	10,679	4,195	−,301	,006
Power	,032	,635	1	19	,435	22,667	,035		
Exponential	,013	,249	1	19	,624	25,281	−,002		

The independent variable is interventions/doctor

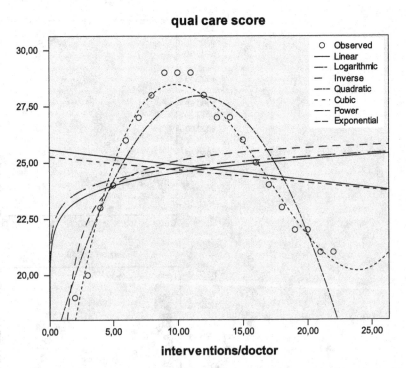

Fig. 11.2 Standard models of regression analyses: the effect of quantity of care (numbers of daily interventions, like endoscopies or small operations, per doctor) is assessed against quality of care scores

4   Logit and Probit Transformations

If linear regression produces a non-significant effect, then other regression functions can be chosen and may provide a better fit for your data. Following logit (= logistic) transformation a linear model is often produced. Logistic regression (odds ratio analysis), Cox regression (Kaplan-Meier curve analysis), Poisson regression (event rate analysis), Markov modeling (survival estimation) are examples. SPSS statistical software [4] covers most of these methods, e.g., in its module "Generalized linear methods". There are examples of datasets where we have prior knowledge that they are linear after a known transformation (Figs. 11.3 and 11.4). As a particular caveat we should add here that many examples can be given, but beware. Most models in biomedicine have considerable residual scatter around the estimated regression line. For example, if the model applied is the following (e = random variation)

$$y_i = \alpha e^{\beta x} + e_i,$$
$$\text{then}$$
$$\ln(y_i) \neq \ln(\alpha) + \beta x + e_i$$

The smaller the e_i term is, the better fit is provided by the model. Another problem with logistic regression is that sometimes after iteration (= computer program for finding the largest log likelihood ratio for fitting the data) the results do not converge, i.e., a best log likelihood ratio is not established. This is due to insufficient data size, inadequate data, or non-quadratic data patterns. An alternative for that purpose

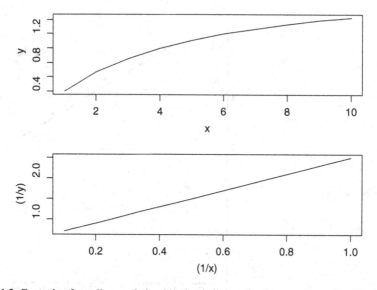

Fig. 11.3 Example of non-linear relationship that is linear after log transformation (Michaelis-Menten relationship between sucrose concentration on x-axis and invertase reaction rate on y-axis)

Fig. 11.4 Another example
of a non-linear relationship
that is linear after logarithmic
transformation (survival of
240 small cell carcinoma
patients)

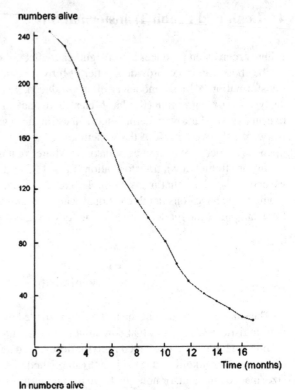

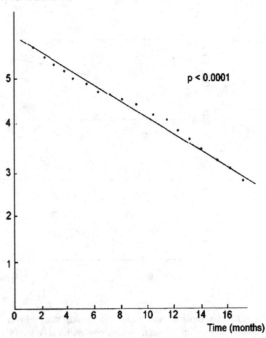

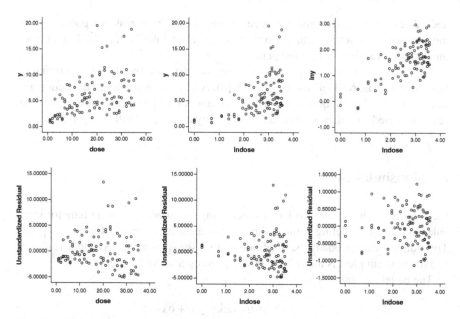

Fig. 11.5 Trial and error methods used to find recognizable data patterns: relationship between isoproterenol dosages (on the x-axis) and relaxation of bronchial smooth muscle (on the y-axis)

is probit modeling, which, generally, gives less iteration problems. The dependent variable of logistic regression (the log odds of responding) is closely related to log probit (probit is the z-value corresponding to its area under curve value of the normal distribution). It can be shown that log odds of responding = logit □ $(\pi/□3)$ × probit. Probit analysis, although not available in SPSS, is in many software programs like, e.g., Stata [5].

5 "Trial and Error" Method, Box Cox Transformation, Ace/Avas Packages

If logit or probit transformations do not work, then additional transformation techniques may be helpful. How do you find the best transformations? First, prior knowledge about the patterns to be expected is helpful. If this is not available, then the "trial and error" method can be recommended, particularly, logarithmically transforming either x- or y-axis or both of them (Fig. 11.5).

$$\log(y) \text{ vs } x, \, y \text{ vs } \log(x), \, \log(y) \text{ vs } \log(x).$$

The above methods can be performed by hand (vs = versus). Box Cox transformation [6], additive regression using ACE [7] (alternating conditional expectations) and AVAS [7] (additive and variance stabilization for regression) packages are modern non-parametric methods, otherwise closely related to the "trial and error" method, can also be used for the purpose.

They are not in SPSS statistical software, but instead a free Box-Cox normality plot calculator is available on the Internet [8]. All of the methods in this section are largely empirical techniques to normalize non-normal data, that can, subsequently, be easily modeled, and they are available in virtually all modern software programs.

6 Sinusoidal Data

Clinical research is often involved in predicting an outcome from a predictor variable, and linear modeling is the commonest and simplest method for that purpose. The simplest except one is the quadratic relationship providing a symmetric curve, and the next simplest is the cubic model providing a sinus-like curve.

The equations are

$$\text{Linear model } y = a + bx$$

$$\text{Quadratic model } y = a + bx^2$$

$$\text{Cubic model } y = a + bx^3$$

The larger the regression coefficient b, the better the model fits the data. Instead of the terms linear, quadratic, and cubic the terms first order, second order, and third order polynomial are applied.

If the data plot looks, obviously, sinusoidal, then higher order polynomial regression and Fourier analysis could be adequate [9]. The equations are given in the appendix. Figure 11.6 gives an example of a polynomial model of the seventh order.

Fig. 11.6 Example of a polynomial regression model of the seventh order to describe ambulatory blood pressure measurements

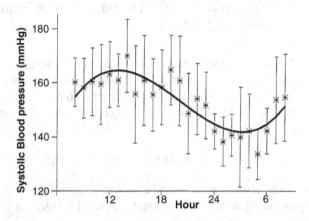

plasmaconcentration

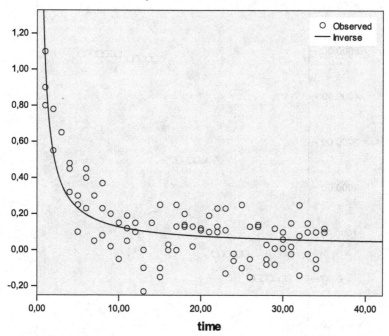

Fig. 11.7 Example of exponential model to describe plasma concentration-time relationship of zoledronic acid

7 Exponential Modeling

For exponential-like patterns like plasma concentration time relationships exponential modeling is a possibility [10]. Also multiple exponential modeling has become possible with the help of Laplace transformations. The non-linear mixed effect exponential model (nonmen model) [11] for pharmacokinetic studies is an example (Fig. 11.7). The data plot shows that the data spread is wide and, so, very accurate predictions can not be made in the given example. Nonetheless, the method is helpful to give an idea about some pharmacokinetic parameters like drug plasma half life and distribution volume.

8 Spline Modeling

If the above models do not adequately fit your data, you may use a method called spline modeling. It stems from the thin flexible wooden splines formerly used by shipbuilders and car designers to produce smooth shapes [1, 2]. Spline modeling

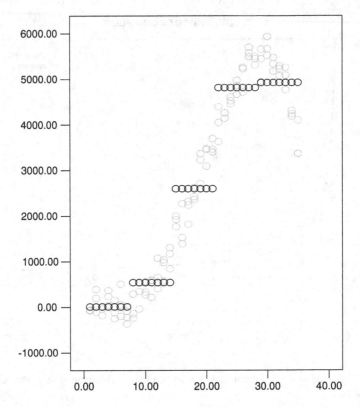

Fig. 11.8 Example of a non-linear dataset suitable for spline modeling: effects of mental stress on fore arm vascular resistance

will be, particularly, suitable for smoothing data patterns, if the data plot leaves you with no idea of the relationship between the y- and x-values.

Figure 11.8 gives an example of non-linear dataset suitable for spline modeling. Technically, the method of local smoothing, categorizing the x-values is used. It means that, if you have no idea about the shape of the relation between the y-values and the x-values of a two dimensional data plot, you may try and divide the x-values into four or five categories, where θ-values are the cut-offs of categories of x-values otherwise called the knots of the spline model.

- cat. 1: min $\leq x < \theta_1$
- cat. 2: $\theta_1 \leq x < \theta_2$
- ---
- cat. k: $\theta_{k-1} \leq x < $ max.

Then, estimate y as the mean of all values within each category. Prerequisites and primary assumptions include

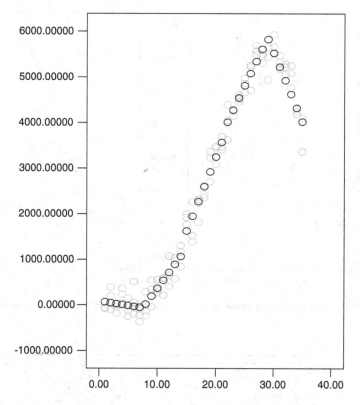

Fig. 11.9 Multiple linear regression lines from the data from Fig. 11.8

- the y-value is more or less constant within categories of the x-values,
- categories should have a decent number of observations,
- preferably, category boundaries should have some meaning.

A linear regression of the categories is possible, but the linear regression lines are not necessarily connected (Fig. 11.9). Instead of linear regression lines a better fit for the data is provided by separate low-order polynomial regression lines (Fig. 11.10). for all of the intervals between two subsequent knots, where knots are x-values that connect one x-category with a subsequent one. Usually, cubic regression, otherwise called third order polynomial regression, is used. It has as simplest equation $y = a + bx^3$. Eventually, the separate lines are joined at the knots. Spline modeling, thus, cuts the data into four or five intervals and uses the best fit third order polynomial functions for each interval (Fig. 11.11). In order to obtain a smooth spline curve the junctions between two subsequent functions must have

1. the same y value,
2. the same slope,
3. the same curvature.

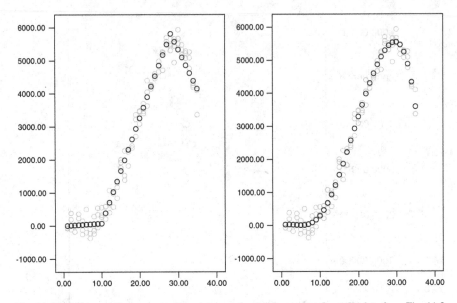

Fig. 11.10 *Left graph* linear regression, *right graph* cubic regression from the data from Fig. 11.8

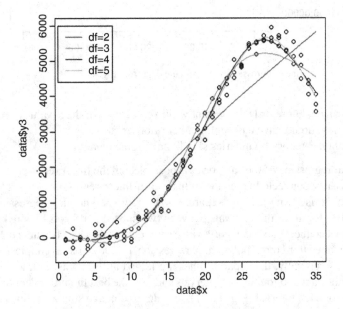

Fig. 11.11 Spline regression of the data from Fig. 11.8 with increasing numbers of knots

All of these requirements are met if

1. the two subsequent functions are equal at the junction,
2. have the same first derivative at the junction,
3. have the same second derivative at the junction.

There is a lot of matrix algebra involved, but a computer program can do the calculations for you, and provide you with the best fit spline curve.

Even with knots as few as 2, cubic spline regression may provide an adequate fit for the data.

In computer graphics spline models are popular curves, because of their accuracy and capacity to fit complex data patterns. So far, they are not yet routinely used in clinical research for making predictions from response patterns, but this is a matter of time. Excel provides free cubic spline function software [12]. The spline model can be checked for its smoothness and fit using lambda-calculus [13], and generalized additive models [14, 15]. Unfortunately, multidimensional smoothing using spline modeling is difficult. Instead you may perform separate procedures for each covariate. Two-dimensional spline modeling is available in SPSS:

Command: graphs....chart builder....basic elements....choose axes....y-x.... gallery....scatter/dot....ok....double click in outcome graph to start chart editor....elements....interpolate....properties....mark: spline....click: apply....best fit spline model is in the outcome graph.

9 Loess Modeling

Maybe, the best fit for many types of non-linear data is offered by still another novel regression method called Loess (locally weighted scatter plot smoothing) [16]. This computationally very intensive program calculates the best fit polynomials from subsets of your data set in order to eventually find out the best fit curve for the overall data set, and is related to Monte Carlo modeling. It does not work with knots, but, instead chooses the bets fit polynomial curve for each value, with outlier values given less weight. Loess modeling is available in SPSS:

Command: graphs....chart builder....basic elements....choose axes....y-x.... gallery....scatter/dot....ok...double click in outcome graph to start chart editor.... elements....fit line at total....properties....mark: Loess....click: apply....best fit Loess model is in the outcome graph.

Figure 11.12 compares the best fit Loess model with the best fit cubic spline model for describing a plasma concentration-time pattern. Both give a better fit for the data than does the traditional exponential modeling with 9 and 29 values in the Loess and spline lines compared to only 5 values in the exponential line of Fig. 11.6. However, it is impossible to estimate plasma half life from Loess and spline. We have to admit that, with so much spread in the data like in the given example, the meaning of the calculated plasma half life is, of course, limited.

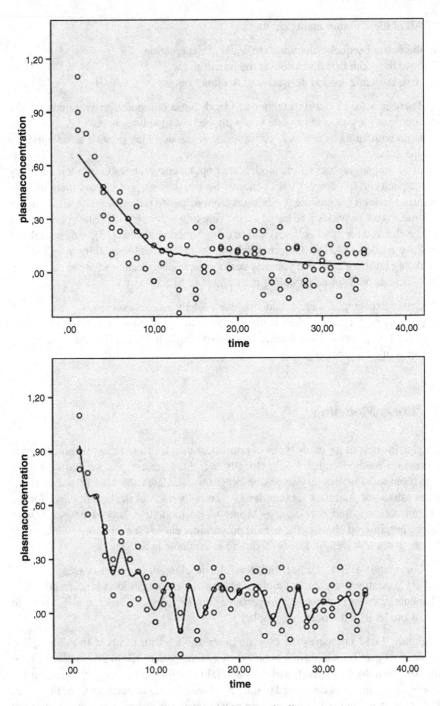

Fig. 11.12 The data from Fig. 11.7 modeled with Loess and spline

10   Discussion

Many tools are available for developing non-linear models for characterizing data sets and making predictions from them. Sometimes it is difficult to choose the degree of smoothness of such models: e.g., with polynomial regression the question is which order, and with spline modeling the questions are how many knots, which locations, which lambdas.

Another method is kernel frequency distribution modeling which unless histograms consists of multiple similarly sized Gaussian curves rather than multiple bins of different length. In order to perform kernel modeling the bandwidth (span) of the Gaussian curves has to be selected which may be a difficult but important factor of the potential fit of a particular kernel method.

Irrespective of the smoothing method applied, there are some problems with smoothing: it may introduce bias, and, second, it may increase the variance in the data. The Akaike information criterion [17] (AIC) is a measure of the relative goodness of fit of a mathematical model for describing data patterns. It can be used to describe the tradeoff between bias and variance in model construction, and to assess the accuracy of the model used. However, the AIC, as it is a relative measure, will not be helpful to confirm a poor result, if all of the models fit the data equally poorly.

Disadvantages of computationally intensive methods like spline modeling and Loess modeling must be mentioned. They require fairly large, densely sampled data sets in order to produce good models. However, the analysis is straightforward. Another disadvantage is the fact that these methods do not produce simple regression functions that can be easily represented by mathematical equations. However, for making predictions from such models direct interpolations/extrapolations from the graphs can be made, and, given the mathematical refinement of these methods, these predictions should, generally, give excellent precision.

11   Conclusions

1. Logit and probit transformation can sometimes be used to mimic a linear model. Logistic regression, Cox regression, Poisson regression, and Markow modeling are examples of logit transformation.
2. Either the x- or y-axis or both of them can be logarithmically transformed. Also Box Cox transformation equation and ace (alternating conditional expectations) or avas (additive and variance stabilization for regression) packages are simple empirical methods often successful for linearly remodeling non linear data.
3. Data that are, obviously, sinusoidal, can, generally, be successfully modeled using polynomial regression and Fourier analysis.
4. For exponential patterns like plasma concentration time relationships exponential modeling with or without Laplace transformations is a possibility.

5. Spline and Loess modeling are modern methods, particularly, suitable for smoothing data patterns, if the data plot leaves you with no idea of the relationship between the y- and x-values. Loess tends to skip outlier data, while spline modeling rather tends to include them. So, if you are planning to investigate the outliers, then spline is your tool.

We have to add that traditional non-linear modeling produces p-values, and modern methods do not. However, given the poor fit of many traditional models, these p-values do not mean too much. Also, it is reassuring to observe that both Loess and spline provide a better fit to non-linear data than does traditional modeling.

12   Appendix

In this appendix the mathematical equations of the non linear models as reviewed are given. They are, particularly, helpful for those trying to understand the assumed relationships between the dependent (y) and independent (x) variables (ln=natural logarithm).

$y = a + b_1 x_1 + b_2 x_2 + \ldots b_{10} x_{10}$	Linear
$y = a + bx + cx^2 + dx^3 + ex^4 \ldots$	Polynomial
$y = a + \text{sinus } x + \text{cosinus } x + \ldots$	Fourier
$\text{Ln odds} = a + b_1 x_1 + b_2 x_2 + \ldots b_{10} x_{10}$	Logistic
Instead of ln odds (= logit) also probit ($\approx \pi\sqrt{3} \times$ logit) is often used for transforming binomial data.	Probit
$\text{Ln multinomial odds} = a + b_1 x_1 + b_2 x_2 + \ldots b_{10} x_{10}$	Multinomial logistic
$\text{Ln hazard} = a + b_1 x_1 + b_2 x_2 + \ldots b_{10} x_{10}$	Cox
$\text{Ln rate} = a + b_1 x_1 + b_2 x_2 + \ldots b_{10} x_{10}$	Poisson
$\log y = a + b_1 x_1 + b_2 x_2 + \ldots b_{10} x_{10}$	Logarithmic
$y = a + b_1 \log x_1 + b_2 x_2 + \ldots b_{10} x_{10}$ etc	"Trial and error"
Transformation function of $y = (y^\lambda - 1) / \lambda$	Box-Cox with λ as power parameter
$y = \left(\text{above transformation function}\right)^{-1}$	ACE modeling
$y = e^{x_1 \ x_2 \ \sin x_3} \ldots \text{etc}$	AVAS modeling
$y = a + e^{b_1 \ x_1} + e^{b_2 \ x_2}$	Multi-exponential modeling
$\theta = $ magnitude of x-value (example)	
$\theta_1 < x < \theta_2$ $\qquad\qquad y = a_1 + b_1 x^3$	Spline modeling
$\theta_2 < x < \theta_3$ $\qquad\qquad y = a_2 + b_2 x^3$	
$\theta_3 < x < \theta_4$ $\qquad\qquad y = a_3 + b_3 x^3$	

References

1. Ferguson JC (1964) Multi-variable curve interpolation. JACM 11:221–228
2. Birkhof F, De Boor R (1964) Piecewise polynomial interpretation and approximation. In: Proceedings of general motors symposium of 1964. Detroit, MI, pp 164–190
3. Cleophas TJ, Zwinderman AH (2009) Monte Carlo methods. In: Statistics applied to clinical trials, 4th edn. Springer, Dordrecht, pp 479–485
4. SPSS Statistical Software. www.spss.com. Accessed 3 Feb 2012
5. Stata Statistical Software. www.stat.com. Accessed 3 Feb 2012
6. Box-Cox normality plot. http://itl.nist.gov/div898/handbook/eda/section3/eda336.htm. Accessed 18 Dec 2012
7. Additive regression and transformation using ace or avas. http://pinard.progiciels-bpi.ca/LibR/library/Hmisc/html/transace.html. Accessed 18 Dec 2012
8. Anonymous. Free statistics and forecasting software, Box-Cox Normality Plot Calculator. www.essa.net/rwasp_boxcoxnorm.wasp/. Accessed 3 Feb 2012
9. Cleophas TJ, Zwinderman AH (2009) Curvilinear regression. In: Statistics applied to clinical trials, 4th edn. Springer, Dordrecht, pp 185–196
10. Cleophas TJ, Zwinderman AH (2009) Regression analysis with Laplace transformations. In: Statistics applied to clinical trials, 4th edn. Springer, Dordrecht, pp 213–216
11. Sheiner LB, Beal SL (1983) Evaluation of methods for estimation of population pharmacokinetic parameters. J Pharmacokinet Pharmacodyn 11:303–319
12. Cubic Spline for Excel. www.srs1software.com/download.htm#pline. Accessed 18 Dec 2012
13. Lambda-calculus. http://en.wikipedia.org/wiki/lambda_calculus. Accessed 18 Dec 2012
14. Hastie T, Tibshirani R (1990) Generalized additive models. Chapman & Hall, London
15. Generalized additive model. http://en.wikipedia.org/wiki/generalized_additive_model. Accessed 18 Dec 2012
16. Local regression. http://en.wikidepia.org/wiki/Local_regression. Accessed 18 Dec 2012
17. Akaike information criterion. http://en.wikipedia.org/wiki/Akaike_information_criterion. Accessed 18 Dec 2012

Chapter 12
Artificial Intelligence, Multilayer Perceptron Modeling

1 Summary

1.1 Background

Back propagation (BP) artificial neural networks is a distribution-free method for data-analysis based on layers of artificial neurons that transduce imputed information. It has been recognized to have a number of advantages compared to traditional methods including the possibility to process imperfect data, and complex nonlinear data.

1.2 Objective

This chapter reviews the principles, procedures, and limitations of BP artificial neural networks for a non-mathematical readership

1.3 Methods and Results

A real data sample of 90 persons' weights, heights and measured body surfaces was used as an example. SPSS 17.0 with neural network add-on was used for the analysis. The predicted body surfaces from a two hidden layer BP neural network were compared to the body surfaces calculated by the Haycock equation. Both the predicted values from the neural network and from the Haycock equation were close to the measured values. A linear regression analysis with neural network as predictor produced an r-square value of 0.983, while the Haycock equation produced a value of 0.995 (r-square > 0.95 is a criterion for accurate diagnostic-testing).

1.4   Conclusions

BP neural networks may, sometimes, predict clinical diagnoses with accuracies similar to those of other methods. However, traditional statistical procedures like regression analyses have to be added for testing their accuracies against alternative methods. Nonetheless, BP neural networks has great potential through its ability to learn by example instead of learning by theory.

2   Introduction

Artificial intelligence is an engineering method that simulates the structures and operating principles of the human brain. Much is unknown of how the brain trains itself to process information, but we do know that brain cells, called neurons, can be activated to send an electric signal through long thin stands called axons. At the end of the axon a structure called the synapse connects the axon with a connected neuron, and provides it with excitatory/inhibitory imput or when the signal is too weak no imput at all. Learning processes in the brain is thought to take place by repeated similar electric signals at similar places giving rise to similar outcomes observed by the brain. This principle can be modeled by artificial neural networks software using observed variables as artificial signals. Software is available in SPSS, MATLAB and so forth: in the current paper SPSS version 17.0, with neural network add-on has been applied [1]. Artificial neural networks are different from traditional statistics that usually assumes Gaussian curve distributions for making predictions from the data. In practice data, sometimes, do not follow Gaussian distributions, and, for that purpose, distribution-free methods, like non-parametric tests and Monte Carlo methods, have been developed. The artificial neural network is another distribution-free method based on layers of artificial neurons that transduce imputed information. It has been recognized to have a number of advantages including the possibility to process imperfect data, and complex non linear data [2]. The current chapter reviews the principles, procedures, and limitations of BP artificial neural networks for a non-mathematical readership.

3   Historical Background

Artificial intelligence was first proposed by the group of neurophysiologist McCulloch in the 1950s [3]. Initially, it was merely to explore and simulate informational processing of the human brain. In the 1960s Rosenblatt [4] developed a three-layer perceptron model (Fig. 12.1), that, with the help of a traditional digital computer system, was capable to process experimental data samples. In the mid-1970s Minsky [5] showed that models with more than three layers were generally required to perform with the precision of current multiple regression models.

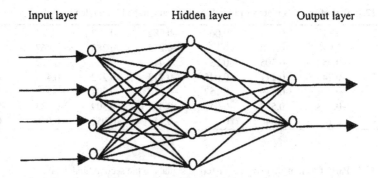

Fig. 12.1 A simple three-layer neural network: each layer of neurons after having received a signal beyond some threshold propagates it forward to the next layer

Particularly, perceptrons with learning samples, otherwise called back propagation (BP) models [6], have been successful in the past two decades, and have been applied for various purposes including sales forecasting, process control, and target marketing [2]. Also in clinical research it has been increasingly applied. In oncology research it has been used for diagnostic purposes [7–14] and survival analysis [15–18], in critical care medicine for patient monitoring [19–23], and in cardiovascular medicine for making diagnoses, including the presence of myocardial infarction [24–26] and coronary artery disease [27–29], and cardiovascular risk predictions [30]. Also in laboratory medicine journals BP neural networks have been published [31, 32].

4   The Back Propagation (BP) Neural Network: The Computer Teaches Itself to Make Predictions

The BP neural networks software include one imput layer, one or more hidden layers and one output layer. Each layer consists of various artificial neurons taking on two phases: activity or inactivity. Figure 12.1 gives a simple example with a single hidden layer. Each neuron in the imput layer after having received a signal beyond some threshold propagates it forward to the next layer. This process will not stop until the signal reaches the output layer sending out the processed signal. The magnitude of the imput values and output values is determined by the structure and functioning of the network. The network is also provided with previously observed outcome data, the so called learning sample. The computer will find, by modifying the weights for all signal-transfers, an outcome as close to the observed outcome as possible. In other words, the neural network tries to find the best-fit outcome for making predictions about the observed outcome data from the imputed data, in a way, similar to regression models. However, unlike regression models no

Table 12.1 Part of weights matrix of transferred signals to the first hidden layer

-0.040	0.370	0.117	0.066	-0.082	-0.227	0.36	-0.321
-0.288	0.070	0.178	-0.190	-0.275	0.283	-0.467	0.032
-0.128	-0.052	-0.305	-0.237	0.442	0.350	0.077	-0.378
-0.585	-0.247	0.271	-0.045	-0.213	0.272	0.403	0.383
0.248	-0.221	-0.149	0.152	-0.012	0.204	-0.233	-0.007
-0.108	-0.338	0.523	-0.046	-0.321	0.309	0.433	-0.068
0.068	0.142	-0.346	0.014	-0.154	-0.052	-0.048	0.160

Table 12.2 Part of weights matrix of transferred signals to the second hidden layer

-0.148	0.602	-0.571	-0.207	0.256	-0.495
0.098	0.684	-0.559	-0.731	1.364	0.097
0.336	-0.541	0.505	-0.241	0.632	-0.188
-0.287	-0.108	0.186	0.124	-0.458	-0.215
0.710	-0.002	-0.387	-0.301	0.735	-0.500

Gaussian distribution models are required, but rather weighted signal-transfers from one layer to another. The Tables 12.1 and 12.2 give examples of weights matrices of imputed signals in the first and second hidden layer of a real data example which will be used in the next section. For finding the best-fit weights the computer uses a technique called iteration or bootstrapping, which means it makes maximally 2,000 guesses, depending on the setting when running the neural network, and then picks out the combination of guesses with the best-fit. The output activity is determined by all imput activities times their weights, and, subsequently, the various hidden layer activities times their weights. The BP principle is, that all imput produces error. Error is assessed, in the usual way, by taking the sums of squared difference from the means of the observed variables. The result with the smallest error is the one with the best-fit. At present, there is no matured theory on how to select the numbers of artificial neurons and hidden layers. The precision of the neural network is improved by feedback signaling (negative weights in the matrices).

5   A Real Data Example

Body surface area is a better indicator for metabolic body mass than body weight, because it is less affected by adipose mass. In laboratory medicine it is used for adjusting oxygen, CO_2 transport parameters, blood volumes, urine creatinine clearance, protein/creatinine ratios and other parameters. The predicting factors of body surface consist of gender, age, weight and height. The body surfaces of 90 persons (Table 12.3) were calculated using direct photometric measurements [33].

The Fig. 12.2 shows the nonlinear relationship between the weights, heights, and measured body surfaces. Using SPSS 17.0 with the neural network add-on module,

Table 12.3 Ninety persons' physical measurements and body surfaces (one row is one person) predicted by mathematical equation and the best-fit result from a two hidden layer neural network

Gender	Age	Weight	Height	Body surface measured	Predicted from equation	Predicted from neural network
var 1	var 2	var 3	var 4	var 5		
1,00	13,00	3,50	138,50	10072,90	10,77,00	10129,64
0,00	5,00	15,00	101,00	6189,00	6490,00	6307,14
0,00	0,00	2,50	51,50	1906,20	1890,00	2565,16
1,00	11,00	30,00	141,00	10290,60	10750,00	10598,32
1,00	15,00	40,50	154,00	13221,60	13080,00	13688,06
0,00	11,00	27,00	136,00	9654,50	10001,00	9682,47
0,00	5,00	15,00	106,00	6768,20	6610,00	6758,45
1,00	5,00	15,00	103,00	6194,10	6540,00	6533,28
1,00	3,00	13,50	96,00	5830,20	6010,00	6096,53
0,00	13,00	36,00	150,00	11759,00	12150,00	11788,01
0,00	3,00	12,00	92,00	5299,40	5540,00	535,63
1,00	0,00	2,50	51,00	2094,50	1890,00	2342,85
0,00	7,00	19,00	121,00	7490,80	7910,00	7815,05
1,00	13,00	28,00	130,50	9521,70	10040,00	9505,63
1,00	0,00	3,00	54,00	2446,20	2130,00	2696,17
0,00	0,00	3,00	51,00	1632,50	2080,00	2345,39
0,00	7,00	21,00	123,00	7958,80	8400,00	7207,74
1,00	11,00	31,00	139,00	10580,80	10880,00	8705,10
1,00	7,00	24,50	122,50	8756,10	9120,00	7978.52
1,00	11,00	26,00	133,00	9573,00	9720,00	9641,04
0,00	9,00	24,50	130,00	9028,00	9330,00	9003,97
1,00	9,00	25,00	124,00	8854,50	9260,00	8804,45
1,00	0,00	2,25	50,50	1928,40	1780,00	2655,69
0,00	11,00	27,00	129,00	9203,10	9800,00	9982,77
0,00	0,00	2,25	53,00	2200,20	1810,00	2582,61
0,00	5,00	16,00	105,00	6785,10	6820,00	7017,29
0,00	9,00	30,00	133,00	10120,80	10500,00	9762,62
0,00	13,00	34,00	148,00	11397,30	11720,00	12063,78
1,00	3,00	16,00	99,00	6410,60	6660,00	6370,21
1,00	3,00	11,00	92,00	5283,30	5290,00	5372,90
0,00	9,00	23,00	126,00	8693,50	8910,00	8450,32
1,00	13,00	30,00	138,00	9626,10	10660,00	11196,58
1,00	9,00	29,00	138,00	10178,70	10460,00	10445,87
1,00	1,00	8,00	76,00	4134,50	4130,00	3952,50
0,00	15,00	42,00	165,00	13019,50	13710,00	13056,80
1,00	15,00	40,00	151,00	12297,10	12890,00	12094,26
1,00	1,00	9,00	80,00	4078,40	4490,00	4520,18
1,00	7,00	22,00	123,00	8651,10	8620,00	8423,78
0,00	1,00	9,50	77,00	4246,10	4560,00	3750,54
1,00	7,00	25,00	125,00	8754,40	9290,00	8398,58
1,00	13,00	36,00	143,00	11282,40	11920,00	11104,75
1,00	3,00	15,00	94,00	6101,60	6300,00	6210,85

(continued)

Table 12.3 (continued)

Gender	Age	Weight	Height	Body surface measured	Predicted from equation	Predicted from neural network
var 1	var 2	var 3	var 4	var 5		
0,00	0,00	3,00	51,00	1850,30	2080,00	2345,39
0,00	1,00	9,00	74,00	3358,50	4360,00	3788,70
0,00	1,00	7,50	73,00	3809,70	3930,00	3800,02
0,00	15,00	43,00	152,00	12998,70	13440,00	13353,48
0,00	13,00	27,50	139,00	9569,10	10200,00	9395,76
0,00	3,00	12,00	91,00	5358,40	5520,00	6090,37
0,00	15,00	40,50	153,00	12627,40	13050,00	12622,94
1,00	5,00	15,00	100,00	6364,50	6460,00	6269,19
1,00	1,00	9,00	80,00	4380,80	4490,00	4520,18
1,00	5,00	16,50	112,00	7256,40	7110,00	7430,72
0,00	3,00	12,50	91,00	5291,50	5640,00	5487,65
1,00	0,00	3,50	56,50	2506,70	2360,00	3065,52
0,00	1,00	10,00	77,00	4180,40	4680,00	3914,55
1,00	9,00	25,00	126,00	8813,70	9320,00	8127,39
1,00	9,00	33,00	138,00	11055,40	11220,00	10561,80
1,00	5,00	16,00	108,00	6988,00	6900,00	6413,58
0,00	11,00	29,00	127,00	9969,80	10130,00	9471,79
0,00	7,00	20,00	114,00	7432,80	7940,00	7299,95
0,00	1,00	7,50	77,00	3934,00	4010,00	4042,95
1,00	11,00	29,50	134,50	9970,50	10450,00	10408,70
0,00	5,00	15,00	101,00	6225,70	6490,00	6307,14
0,00	3,00	13,00	91,00	5601,70	5760,00	5623,51
0,00	5,00	15,00	98,00	6163,70	6410,00	6296,79
1,00	15,00	45,00	157,00	13426,70	13950,00	13877,81
1,00	7,00	21,00	120,00	8249,20	8320,00	8445,74
0,00	9,00	23,00	127,00	8875,80	8940,00	9023,25
0,00	7,00	17,00	104,00	6873,50	7020,00	6935,27
1,00	15,00	43,50	150,00	13082,80	13450,00	13508,38
1,00	15,00	50,00	168,00	14832,00	15160,00	13541,31
0,00	7,00	18,00	114,00	7071,80	7510,00	7161,82
1,00	3,00	14,00	97,00	6013,60	6150,00	6200,79
1,00	7,00	20,00	119,00	7876,40	8080,00	7606,17
0,00	0,00	3,00	54,00	2117,30	2130,00	2559,28
1,00	1,00	9,50	74,00	4314,20	4490,00	4531,14
0,00	15,00	44,00	163,00	13480,90	13990,00	13612,74
0,00	11,00	32,00	140,00	10583,80	11100,00	10401,88
1,00	0,00	3,00	52,00	2121,00	2100,00	2337,69
0,00	11,00	29,00	141,00	10135,30	10550,00	10291,93
0,00	3,00	15,00	94,00	6074,90	6300,00	6440,60
0,00	13,00	44,00	140,00	13020,30	13170,00	12521,73
1,00	5,00	15,50	105,00	6406,50	6700,00	6532,15
1,00	9,00	22,00	126,00	8267,00	8700,00	8056,85
0,00	15,00	40,00	159,50	12769,70	13170,00	12994,08

(continued)

Table 12.3 (continued)

Gender	Age	Weight	Height	Body surface measured	Predicted from equation	Predicted from neural network
var 1	var 2	var 3	var 4	var 5		
1,00	1,00	9,50	76,00	3845,90	4530,00	4240,36
0,00	13,00	32,00	144,00	10822,10	11220,00	10964,35
1,00	13,00	40,00	151,00	12519,90	12890,00	12045,33
0,00	9,00	22,00	124,00	8586,10	8650,00	8411,62
1,00	11,00	31,00	135,00	10120,60	10750,00	9934,60

The best-fit result as presented had an error as small as 0.0035 obtained after maximally 2,000 iterations
Var means variable

Fig. 12.2 VAR00003 = weight; VAR00004 = height; VAR00005 = measured body surface. The three dimensional scatter plot shows the nonlinear relationship between the variables

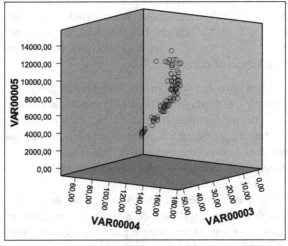

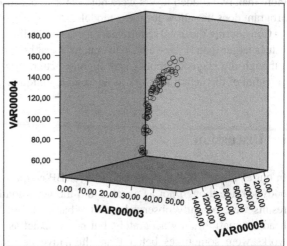

we assess whether a neural network with two hidden layers would be able to adequately predict the measured body surfaces, and whether it would perform better than the Haycock equation (* = sign of multiplication) [34]:

$$\text{body surface} = 0.024265 * \text{height}^{0.3964} * \text{weight}^{0.5378}.$$

The data file consists of a row for each person with different factors and one dependent variable, the measured body surface (Table 12.3). We command: neural networks; multilayer perceptron. Select the dependent variable, the measured body surface, factors, body height and weight, and covariates, age and gender, in the main dialog box. Here are also various dialog boxes that can be assessed from the main dialog box:

1. the dialog box partitioning: set the training sample (70), test sample (20)
2.   "   "   architecture: set the numbers of hidden layers (2)
3.   "   "   activation function: click hyperbolic tangens
4.   "   "   output: click diagrams, descriptions, synaptic weights
5.   "   "   training: maximal time for calculations 15 min, maximal numbers of iterations 2,000.

Then press ok, and synaptic weights and body surfaces predicted by the neural network are displayed as well as the smallest error. The results are in Table 12.3. Also, the values obtained from the Haycock equation are included in the table.

Both the predicted values from the neural network and from the Haycock equation are close to the measured values. When performing a linear regression with neural network as predictor, the r-square value was 0.983, while the Haycock produced an r-square value of 0.995.

In order to assess the robustness of the neural network result, smaller training samples, fewer iterations and a single hidden layer were assessed, but all of these changes produced smaller r-square values indicating less precision. Models with more than two hidden layers were not assessed, because the SPSS software add-on program does not allow for these models.

Considering the usual requirement for accurate diagnostic testing [35] of r-square values larger than 0.95, we have to conclude that both methods perform adequately, although the Haycock model was slightly better. The example still illustrates the potential of the neural network as an exact technique for predicting body surfaces.

6   Discussion

In our example of 90 persons a four layer BP neural network accurately predicted body surface though not as exact as did the usual mathematical equation. Similar results were recently observed by Eftekbar et al. [23]: neural network predicted head trauma mortality accurately, but not as exact as logistic models. Neural networks were sometimes better than alternative procedures. For example, in 331 adults with chest pain for making a diagnosis of acute infarction they were better than were the attending emergency room physicians (sensitivity and specificity 97 and 96 versus 78 and 85%) [25]. Also, in 1,107 patients BP neural networks was more sensitive to detect anterior and inferior infarctions than conventional automated electrocardiogram interpretation (81 and 78 versus 68 and 66%) [36]. Nowadays, intelligent computing techniques, mimicking the brain, receive a great deal of attention from the scientific world: e.g., the Google database system gives approximately ten million hits for the search term "artificial intelligence". Yet, many questions are unanswered. For example, artificial intelligence does not possess human brain-characteristics like tolerance, robustness, and levels of consciousness. Also, methods that are more prone to generalization, such as the BP based neural network techniques and their computational systems are generally unable to provide a definite explication of the outcome, sometimes leading to incorrect conclusions, and, if used for the classifications of single cases, to questionable results.

We should add that data driven analyses are, generally, not a sound basis for scientific research, and a major source of misunderstandings due to results based on chance rather than true effects. Neural network by its very nature of black box modeling is at risk of being abused for that purpose. When applied for clinical research, particularly diagnostic research, it should be based on appropriate prior hypotheses and prior knowledge. Fortunately, this has been recognized and emphasized by several investigators [37–40].

Regarding the potential users of neural network methods, we believe that, despite the requirement of basic statistical knowledge, current user-friendly statistical software like the SPSS add-on module "Neural Network" [1] can be used by clinical and laboratory investigators without the help of a statistician. After all, those who invented artificial intelligence were neurologists and neurophysiologists, rather than statisticians [2].

Although artificial intelligence may *approximately* correspond to the intelligence of human beings, it is, probably, also largely different from the brain, and should, currently, be interpreted as just another non-Gaussian method for data assessment. Moreover, its mathematical basis is not fully recognized. Also, traditional statistical methods like regression methods have to be added for testing its accuracy against alternative methods. Nonetheless, it has great potential through its ability to learn by example instead of learning by theory, making it very flexible and powerful.

7 Conclusions

Back propagation (BP) artificial neural networks is a distribution-free method for data-analysis based on layers of artificial neurons that transduce imputed information. It has been recognized to have a number of advantages compared to traditional methods including the possibility to process imperfect data, and complex nonlinear data. This chapter reviews the principles, procedures, and limitations of BP artificial neural networks for a non-mathematical readership

A real data sample of 90 persons' weights, heights and measured body surfaces was used as an example. SPSS 17.0 with neural network add-on was used for the analysis. The predicted body surfaces from a two hidden layer BP neural network were compared to the body surfaces calculated by the Haycock equation. Both the predicted values from the neural network and from the Haycock equation were close to the measured values. A linear regression analysis with neural network as predictor produced an r-square value of 0.983, while the Haycock equation produced a value of 0.995 (r-square>0.95 is a criterion for accurate diagnostic-testing).

BP neural networks may, sometimes, predict clinical diagnoses with accuracies similar to those of other methods. However, traditional statistical procedures like regression analyses have to be added for testing their accuracies against alternative methods. Nonetheless, BP neural networks has great potential through its ability to learn by example instead of learning by theory.

References

1. WWW.SPSS.COM
2. Stergiou C, Siganos D. Neural networks. www.doc.ic.ac.uk
3. Andrew AM (2004) Work of Warren McCulloch. Kybernetes 33:141–146
4. Rosenblatt F (1962) Principles of neurodynamics: perceptrons and the theory of brain mechanisms. Spartan, New York
5. Minsky MA (1974) Framework for representing knowledge. Technical report Massachusetts Institute of Technology, AIM-306, Cambridge, MA, USA
6. Rumbelhart DE, Hinton GE, Williams RJ (1986) Learning representations by back -propagating errors. Nature 323:533–536
7. Simpson JH, McArdle C, Pauson AW, Hume P, Turkes A, Griffiths K (1995) A non -invasive test for the pre-cancerous breast. Eur J Cancer 31A:1768–1772
8. Naguib RN, Adams AE, Horne CH, Angus B, Sherbet GV, Lennard TW (1996) The detection of nodal metastasis in breast cancer using neural networks. Physiol Meas 17:297–303
9. Sherman ME, Schiffman MH, Mango LJ, Kelly D, Acosta D, Cason Z, Elgert P, Zaleski S, Scot DR, Kurman R, Stoler M, Lorincz AT (1997) Evaluation of PAPNET testing as an ancillary tool to clarify the status of the atypical cervical smear. Mod Pathol 10:564–567
10. Mango LJ, Valente PT (1998) Neural networks assisted analysis and microscopic rescreening in presumed negative cervical cytologic smears. Acta Cytol 42:227–232
11. Doornewaard H, Van der Schouw YT, Van der Graaf Y, Bos AB, Habbema JD, Van den Tweel JG (1999) The diagnostic value of computer assisted primary smear screening: a longitudinal cohort study. Mod Pathol 12:995–1000
12. Prismatic Project Management Team (1999) Assessment of automated primary screening on PAPNET of cervical smears in the PRISMATIC trial. Lancet 353:1381–1385
13. Finne P, Finne R, Auvinen A, Juusela H, Aro J, Maattanen L, Hakama M, Ranniko S, Tammela TL, Stenman U (2000) Predicting the outcome of prostate biopsy in screen positive men by a multilayer perceptron network. Urology 56:418–422
14. Gamito EJ, Stone NN, Batuello JT, Crawford ED (2000) Use of artificial neural networks in the clinical staging of prostate cancer. Tech Urol 6:60–63
15. Bugliosi R, Tribalto M, Avvisati G, Boccardoro M, De Martinis C, Friera R, Mandelli F, Pileri A, Papa G (1994) Classification of patients affected by multiple myeloma using neural network software. Eur J Haematol 52:182–183

16. Kothari R, Cualing H, Balachander T (1996) Neural network analysis of flow cytometry immunophenotype data. IEEE Biomed Eng 43:803–810

17. Glas JO, Reddick WE (1998) Hybrid artificial neural network segmentation and classification of dynamic contrast enhanced MR imaging of osteosarcoma. Magn Reson Imaging 16:1075–1083

18. Bryce TJ, Dewhirst MW, Floyd CE, Hars V, Brizel DM (1998) Artificial neural networks of survival in patients treated with irradiation with and without concurrent chemotherapy for advanced carcinoma of the head and neck. Int J Radiat Oncol Biol Phys 41:339–345

19. Stock A, Rogers MS, Li A, Chang AM (1994) Use of neural networks for hypothesis generation in fetal surveillance. Baillieres Clin Obstet Gynaecol 8:533–548

20. Si Y, Gotman J, Pasupathy A, Flanagan D, Rosenblatt B, Gottesman R (1998) An expert system for EEG monitoring in the pediatric intensive care. Electroencephalogr Clin Neurophysiol 106:488–500

21. Zernikow B, Holtmannspotter K, Michel E, Theilhaber M, Pielemeier W, Hennecke KH (1998) Artificial neural network for predicting intracranial haemorrhage in preterm neonates. Acta Paediatr 87:969–975

22. Zernikow B, Holtmannspotter K, Michel E, Hornschuh F, Groote K, Hennecke KH (1999) Predicting length of stay in preterm neonates. Eur J Pediatr 158:59–62

23. Eftekbar B, Mohammad K, Ardebilli HE, Ghodsi M, Ketabchi E (2005) Comparison of artificial neural network and regression models for prediction of mortality in head trauma based on clinical data. BMC Med Inf Decis Mak 5:3–9

24. Selker HP, Griffith JL, Patil S, Long WJ, D'Agostino RB (1995) A comparison of performance of mathematical predictive methods for medical diagnosis: identifying acute cardiac ischemia among emergency department patients. J Investig Med 43:468–476

25. Baxt WG, Skora J (1996) Prospective validation of artificial neural network trained to identify acute myocardial infarction. Lancet 347:12–15

26. Ellenius J, Groth T, Lindahl B (1997) Neural network of biochemical markers for early assessment of acute myocardial infarction. Stud Health Technol Inform 43:382–385

27. Goodenday LS, Cios KJ, Shin L (1997) Identifying coronary stenosis using an image recognition neural network. IEEE Eng Med Bio Mag 16:139–144

28. Polak MJ, Zhou SH, Rautaharju PM, Armstrong WW, Chaitman BR (1997) Using automated analysis of resting twelve lead ECG to identify patients at risk of developing transient myocardial ischaemia. Physiol Meas 18:317–325

29. Lindahl D, Toft J, Hesse B, Palmer J, Ali S, Lundin A, Edenbrandt L (2000) Scandinavian test of artificial neural network for classification of myocardial perfusion images. Clin Physiol 20:253–261

30. Patil N, Smith TJ. Neural network analysis speeds disease risk predictions, innovative clinical models transform cardiovascular assessment algorithms. In: Scientific computing 2009, Rockaway NJ, p 07866. www.scientificcomputing.com. Accessed 18 Dec 2012

31. Queralto JM, Torres J, Guinot M (1999) Neural networks for the biochemical prediction of bone mass. Clin Chem Lab Med 37:831–838

32. Papik K, Molnar B, Fedorczak P, Schaefer R, Lang F, Sreter L, Feher J, Tulassay Z (1999) Automated prozone effect detection in ferritin homogenous assays using neural networks. Clin Chem Lab Med 37:471–476

33. Mitchell D, Strydom NB, Van Graan CH, Van der Walt H (1971) Human surface area: comparison of the du Bois formula with direct photometric measurement. Eur J Physiol 325:188–190

34. Haycock GB, Schwarz GJ, Wisotsky DH (1978) Body surface area calculated from the height and weight. J Pediatr 93:62–66

35. Atiqi R, Van Iersel C, Cleophas TJ (2009) Accuracy of quantitative diagnostic tests. Int J Clin Pharmacol Ther 47:153–159

36. Heden B, Edenbrandt L, Hasity WK, Pahlm O (1994) Artificial neural networks for electrocardiographic diagnosis of healed myocardial infarction. Am J Cardiol 74:5–8

37. Redding NJ, Kowalczyk A, Downs T (1993) Constructive higher order network algorithms that is polynomial time. Neural Netw 6:997–1010
38. Sperduti A, Starita A (1993) Speed up learning and network optimization with extended back propagation. Neural Netw 6:365–383
39. Wnek J, Michalski RS (1994) Hypothesis driven constructive induction in AQ17-HCI: a method and experiments. Mach Learn 14:139–168
40. Lytton WW (2002) From artificial neural network to realistic neural network, Chapter 14. In: From computer to brain. Springer, New York, pp 259–268

Chapter 13
Artificial Intelligence, Radial Basis Functions

1 Summary

1.1 Background

Radial basis function network may better than multilayer perceptron neural network predict medical data, because it uses a Gaussian activation function, but it is rarely used.

1.2 Objective

To assess the performance of radial basis function networks in clinical research.

1.3 Methods

A 90 person study of the variables age, gender, weight and height to predict body surface was applied as an example.

1.4 Results

The radial basis function predicted the measured values with a correlation coefficient of 0.985 (r-square 0.970). The multilayer perceptron and the Hancock's mathematical equation produced r-square values of respectively 0.983 and 0.995. A trend to skewness of height and a significant kurtosis of age may have slightly

T.J. Cleophas and A.H. Zwinderman, *Machine Learning in Medicine*,
DOI 10.1007/978-94-007-5824-7_13, © Springer Science+Business Media Dordrecht 2013

reduced the performance of the radial basis function in the example given. Yet as r-square values larger than 0.95 indicate excellent accuracy, both types of neural networks performed excellently and nearly as well as the mathematical equation.

1.5 Conclusions

1. Both types of neural networks are currently listed as supervised machine learning methods and rightly so, because they learn computers to make health predictions for the benefit of health in mankind.
2. In our example both types of neural networks performed very well and virtually similarly well as a traditional mathematical equation.

2 Introduction

Radial basis functions are symmetric functions around the origin. These functions are particularly convenient for describing data that are systematically equidistant to an origin. As a matter of fact the equations of Gaussian curves are radial distant functions, and they are particularly useful for making predictions about scattered data as proposed by Richard Franke, mathematician at the Naval Postgraduate School Monterey in 1982 [1] and one of the founders of radial basis networks, a special type of neural network. Neural networks, sometimes called artificial intelligence is an engineering method that simulates the structures and operating principles of the human brain [2]. The multi-layer perceptron neural network software includes one imput layer, at least one hidden layer and one output layer of artificial neurons. Each neuron after having received a signal beyond some threshold propagates it forward to the next layer. Using a learning data sample with measured outcome values the computer can teach itself to make predictions about future data in a way similar to the supposed learning processes in the brain [2]. The activation function for transmission of the signals is traditionally sigmoidal. Radial basis function networks is different from the traditional neural network, because a Gaussian activation function instead of a sigmoidal one is used. This may have advantages, particularly with Gaussian-like predictor data [3].

The current chapter was written to assess the performance of radial basis networks as compared to that of the traditional multilayer methods. Radial basis function networks despite its short history are currently already routinely used in fields like marketing research [4] and geoscience [5], but it is rarely used in medicine despite the omnipresence of Gaussian data in clinical research. Searching Medline we found one gynaecological [6], one cardiovascular [7], one neurological [8], and one infection study [9].

This chapter was written as a hand-hold presentation accessible to clinicians, and as a must-read publication for those new to the method. It is the author's experience, as a master class professor, that students are eager to master adequate command of statistical software. For their benefit all of the steps of the novel method from logging in to the final result using SPSS statistical software [10] will be given.

3   Example

We used the same example as we previously did in a traditional multi-layer perceptron neural network assessment [11]. Body surface area is a better indicator for metabolic body mass than body weight, because it is less affected by adipose mass. In laboratory medicine it is used for adjusting oxygen, CO_2 transport parameters, blood volumes, urine creatinine clearance, protein/creatinine ratios and other parameters. The predicting factors of body surface consist of gender, age, weight and height. The body surfaces of 90 persons (Table 13.1) were calculated using direct photometric measurements [12]. These previously measured outcome data will be used as the so called learning sample, and the computer will be commanded to teach itself making predictions about the body surface from the predictor variables gender, age, weight and height.

We will also assess the performance of the traditional mathematically modeled Haycock's equation (* = sign of multiplication) [13] for calculating body surface:

$$\text{body surface} = 0.024265 * \text{height}^{0.3964} * \text{weight}^{0.5378}.$$

4   Radial Basis Function Analysis

For analysis the SPSS module Neural Networks is used [10].

We command: Analyze....Neural Networks....Radial Basis Function.... Dependent Variables: enter Body surface measured....Factors: enter gender, age, weight, and height....Partitions: Training 7....Test 3....Holdout 0....Output: mark Description....Diagram....Model summary....Predicted by observed chart....Case processing summary....Save: mark Save predicted value of category for each dependent variable....automatically generate unique names....OK.

Figure 13.1 shows the three layer radial basis network with one imput layer, one hidden layer and one outcome layer. Figure 13.2 gives the relationship between the body surface measured versus the body surface predicted by the radial basis function model in 90 persons. The Table 13.1 right end column gives the body surface values predicted by the radial basis network. Some cases in the testing sample had factor values that did not occur in the training sample. These cases were excluded from the analysis. The fifth column of Table 13.1 gives the measured body surface values. In order to assess the correlation between the measured and predicted values, linear regression analysis is performed. Table 13.2 upper part shows that the

Table 13.1 Data file of the example used

Var 1 gender
Var 2 age
Var 3 weight
Var 4 height
Var 5 body surface measured
Var 6 body surface predicted from the Hancock's equation
Var 7 Predicted Value from radial basis function network

Var 1	Var 2	Var 3	Var 4	Var 5	Var 6	Var 7
1,00	13,00	30,50	138,50	10072,90	10770,00	10875,08
0,00	5,00	15,00	101,00	6189,00	6490,00	6455,37
000	0,00	2,50	51,50	1906,20	1890,00	2108,16
1,00	11,00	30,00	141,00	10290,60	10750,00	9996,18
1,00	15,00	40,50	154,00	13221,60	13080,00	13471,14
0,00	11,00	27,00	136,00	9654,50	10000,00	9849,49
0,00	5,00	15,00	106,00	6768,20	6610,00	6455,37
1,00	5,00	15,00	103,00	6194,10	6540,00	6436,05
1,00	3,00	13,50	96,00	5830,20	6010,00	5832,91
0,00	13,00	36,00	150,00	11759,00	12150,00	10952,11
0,00	3,00	12,00	92,00	5299,40	5540,00	5649,20
1,00	0,00	2,50	51,00	2094,50	1890,00	2001,12
0,00	7,00	19,00	121,00	7490,80	7910,00	7898,07
1,00	13,00	28,00	130,50	9521,70	10040,00	10875,08
1,00	0,00	3,00	54,00	2446,20	2130,00	2001,12
0,00	0,00	3,00	51,00	1632,50	2080,00	1852,02
0,00	7,00	21,00	123,00	7958,80	8400,00	7898,07
1,00	11,00	31,00	139,00	10580,80	10880,00	10051,87
1,00	7,00	24,50	122,50	8756,10	9120,00	7931,94
1,00	11,00	26,00	133,00	9573,00	9720,00	9927,07
0,00	9,00	24,50	130,00	9028,00	9330,00	9516,07
1,00	9,00	25,00	124,00	8854,50	9260,00	9532,99
1,00	0,00	2,25	50,50	1928,40	1780,00	2198,58
0,00	11,00	27,00	129,00	9203,10	9800,00	9849,49
0,00	0,00	2,25	53,00	2200,20	1810,00	2108,16
0,00	5,00	16,00	105,00	6785,10	6820,00	6613,17
0,00	9,00	30,00	133,00	10120,80	10500,00	9523,46
0,00	13,00	34,00	148,00	11397,30	11720,00	
1,00	3,00	16,00	99,00	6410,60	6660,00	
1,00	3,00	11,00	92,00	5283,30	5290,00	5692,93
0,00	9,00	23,00	126,00	8693,50	8910,00	
1,00	13,00	30,00	138,00	9626,10	10660,00	10859,53
1,00	9,00	29,00	138,00	10178,70	10460,00	9539,40
1,00	1,00	8,00	76,00	4134,50	4130,00	
0,00	15,00	42,00	165,00	13019,50	13710,00	13451,21
1,00	15,00	40,00	151,00	12297,10	12890,00	12429,90
1,00	1,00	9,00	80,00	4078,40	4490,00	4053,74
1,00	7,00	22,00	123,00	8651,10	8620,00	7931,94
0,00	1,00	9,50	77,00	4246,10	4560,00	3831,38
1,00	7,00	25,00	125,00	8754,40	9290,00	7931,94
1,00	13,00	36,00	143,00	11282,40	11920,00	10872,54
1,00	3,00	15,00	94,00	6101,60	6300,00	5613,32
0,00	0,00	3,00	51,00	1850,30	2080,00	1852,02
0,00	1,00	9,00	74,00	3358,50	4360,00	3892,26

(continued)

Table 13.1 (continued)

Var 1 gender						
0,00	1,00	7,50	73,00	3809,70	3930,00	4151,78
0,00	15,00	43,00	152,00	12998,70	13440,00	
0,00	13,00	27,50	139,00	9569,10	10200,00	10879,05
0,00	3,00	12,00	91,00	5358,40	5520,00	5649,20
0,00	15,00	40,50	153,00	12627,40	13050,00	13491,78
1,00	5,00	15,00	100,00	6364,50	6460,00	6436,05
1,00	1,00	9,00	80,00	4380,80	4490,00	4053,74
1,00	5,00	16,50	112,00	7256,40	7110,00	6865,88
0,00	3,00	12,50	91,00	5291,50	5640,00	5649,20
1,00	0,00	3,50	56,50	2506,70	2360,00	2522,45
0,00	1,00	10,00	77,00	4180,40	4680,00	3991,76
1,00	9,00	25,00	126,00	8813,70	9320,00	
1,00	9,00	33,00	138,00	11055,40	11220,00	9539,13
1,00	5,00	16,00	108,00	6988,00	6900,00	6679,86
0,00	11,00	29,00	127,00	9969,80	10130,00	9861,75
0,00	7,00	20,00	114,00	7432,80	7940,00	7859,87
0,00	1,00	7,50	77,00	3934,00	4010,00	3892,26
1,00	11,00	29,50	134,50	9970,50	10450,00	
0,00	5,00	15,00	101,00	6225,70	6490,00	6455,37
0,00	3,00	13,00	91,00	5601,70	5760,00	5649,20
0,00	5,00	15,00	98,00	6163,70	6410,00	6455,37
1,00	15,00	45,00	157,00	13426,70	13950,00	13419,00
1,00	7,00	21,00	120,00	8249,20	8320,00	7929,16
0,00	9,00	23,00	127,00	8875,80	8940,00	9522,21
0,00	7,00	17,00	104,00	6873,50	7020,00	7898,07
1,00	15,00	43,50	150,00	13082,80	13450,00	13375,48
1,00	15,00	50,00	168,00	14832,00	15160,00	13419,00
0,00	7,00	18,00	114,00	7071,80	7510,00	7859,87
1,00	3,00	14,00	97,00	6013,60	6150,00	5832,91
1,00	7,00	20,00	119,00	7876,40	8080,00	
0,00	0,00	3,00	54,00	2117,30	2130,00	1947,16
1,00	1,00	9,50	74,00	4314,20	4490,00	4053,74
0,00	15,00	44,00	163,00	13480,90	13990,00	13491,78
0,00	11,00	32,00	140,00	10583,80	11100,00	9881,66
1,00	0,00	3,00	52,00	2121,00	2100,00	2001,12
0,00	11,00	29,00	141,00	10135,30	10550,00	9854,36
0,00	3,00	15,00	94,00	6074,90	6300,00	5584,94
0,00	13,00	44,00	140,00	13020,30	13170,00	11890,07
1,00	5,00	15,50	105,00	6406,50	6700,00	6679,86
1,00	9,00	22,00	126,00	8267,00	8700,00	
0,00	15,00	40,00	159,50	12769,70	13170,00	
1,00	1,00	9,50	76,00	3845,90	4530,00	4250,24
0,00	13,00	32,00	144,00	10822,10	11220,00	
1,00	13,00	40,00	151,00	12519,90	12890,00	10871,09
0,00	9,00	22,00	124,00	8586,10	8650,00	9505,96
1,00	11,00	31,00	135,00	10120,60	10750,00	9929,20

Some cases in the testing sample had factor values that did not occur in the training sample. These cases were excluded from the analysis

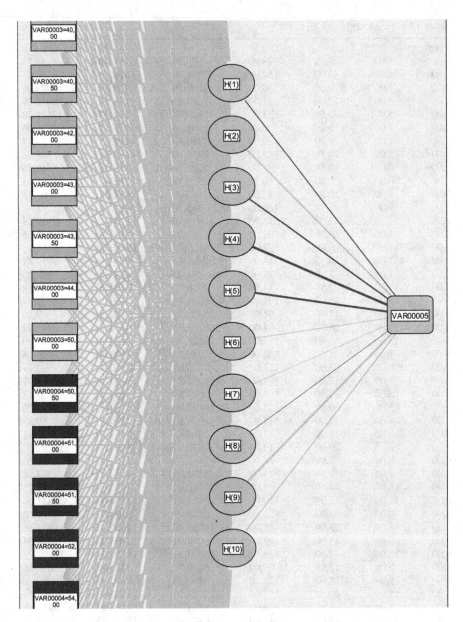

Fig. 13.1 The three layer radial basis network with one imput layer, one hidden layer and one outcome layer (here only one variable)

radial basis function provided very accurate predictions of the measured values with a correlation coefficient close to 1 (namely 0.985, r-square 0.970).

We also used the calculated body surface values from the Hancock's equation (Table 13.1, sixth column). When assessing the correlation of the Hancock's values with the measured values, a somewhat larger correlation coefficient came up

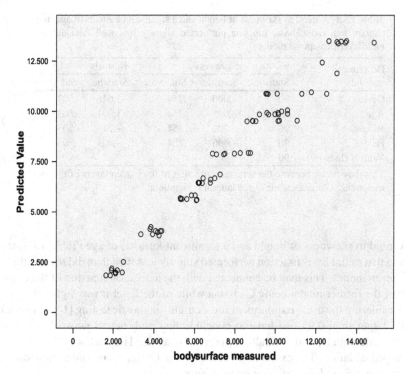

Fig. 13.2 The relationship between the body surface measured versus the body surface predicted by the radial basis function model in 90 persons

Table 13.2 Although the Hancock's equation performed slightly better (lower part), the performance of radial basis function was excellent (upper part)

Model summary				
Model	R	R Square	Adjusted R Square	Std. error of the estimate
1	,985[a]	,970	,970	595,61690
Model summary				
Model	R	R Square	Adjusted R Square	Std. error of the estimate
1	,998[b]	,995	,995	229,38110

We should add that the performance of a more advanced multilayer perceptron neural network provided an r square value of 0.983, and was thus slightly better than the radial basis function network

[a]Predictors: (constant), predicted value for VAR00005
[b]Predictors: (constant), predicted from equation

(Table 13.2, lower part), namely 0.998, r-square 0.995. Although the Hancock's equation performed better, the performance of radial basis function was still excellent. We should add that the performance of a more advanced multilayer perceptron neural network provided an r square value of 0.983, and was thus, slightly, better than the radial basis function network [11].

Table 13.3 A trend to skewness of height and a significant kurtosis of age may explain that radial basis function performed slightly less well than did the multilayer perceptron model

Descriptive statistics	N Statistic	Skewness Statistic	Std. error	Kurtosis Statistic	Std. error
Gender	90	,000	,254	−2,046	,503
Age	90	,085	,254	−1,301	,503
Weight	90	,302	,254	−,772	,503
Height	90	−,406	,254	−,803	,503
Valid N (listwise)	90				

This may be so, because the activation function of the hidden layers of the former model Gaussian while of the latter it is sigmoidal

A trend to skewness of height and a significant kurtosis of age (Table 13.3) may explain that radial basis function performed slightly less well than did the multilayer perceptron model. This may be connected with the activation function of the hidden layer of the former model being Gaussian while of the latter it was sigmoidal.

Considering the usual requirement for accurate diagnostic testing [14] of r-square values larger than 0.95, we have to conclude that both neural networks methods perform adequately, and virtually equally well as the Hancock's mathematically developed equation. The example illustrates the potential of the neural networks as an exact technique for predicting body surfaces.

5   Discussion

In our example of 90 persons a three layer radial basis function neural network accurately predicted body surface though not as exact as did the usual mathematical equation. Also a four layer multilayer perceptron model performed slightly better (though again not as well as the mathematical equation). Similar results were recently observed by Eftekbar et al. [15]: neural network predicted head trauma mortality accurately, but not as exact as logistic models. Neural networks were sometimes better than alternative procedures. For example, in 331 adults with chest pain for making a diagnosis of acute infarction they were better than were the attending emergency room physicians (sensitivity and specificity 97 and 96 versus 78 and 85%) [16]. Also, in 1,107 patients BP neural networks was more sensitive to detect anterior and inferior infarctions than conventional automated electrocardiogram interpretation (81 and 78 versus 68 and 66%) [17]. Nowadays, intelligent computing techniques, mimicking the brain, receive a great deal of attention from the scientific world: e.g., the Google database system gives approximately ten million hits for the search term "artificial intelligence". Yet, many questions are unanswered. For example, artificial intelligence does not possess human brain-characteristics like tolerance, robustness, and levels of consciousness. Also, methods that are more prone to generalization, such as the BP based neural network techniques and their computational systems are generally unable to provide a definite explication of the outcome, sometimes leading to incorrect conclusions, and, if used for the classifications of single cases, to questionable results.

Data driven analyses are, generally, not a sound basis for scientific research, and a major source of misunderstandings due to results based on chance rather than true effects. Neural network by its very nature of black box modeling is at risk of being abused for that purpose. When applied for clinical research, particularly diagnostic research, it should be based on appropriate prior hypotheses and prior knowledge. Fortunately, this has been recognized and emphasized by several investigators [18–21].

We should add that both types of neural networks are currently listed as supervised machine learning methods and rightly so, because they enable computers to teach themselves making health predictions for the benefit of health in mankind.

6 Conclusions

1. Radial basis function networks is different from the traditional neural network, because a Gaussian activation function instead of a sigmoidal one is used. This may have advantages, particularly with Gaussian-like clinical data.
2. In our example of 90 persons a three layer radial basis function neural network accurately predicted body surface though not as exact as did the usual mathematical equation.
3. Also a four layer multilayer perceptron model performed slightly better (though again not as well as the mathematical equation).
4. Both types of neural networks are currently listed as supervised machine learning methods and rightly so, because they learn computers to make health predictions for the benefit of health in mankind.

References

1. Franke R (1982) Scattered data interpolation: tests of some methods. Math Comput 38:181–200
2. Rumelhart DE, Hinton GE, Williams RJ (1986) Learning representations by back propagating errors. Nature 323:533–536
3. Xie T, Yu H, Wilamowski B (2011) Comparison between traditional neural networks, and radial basis function networks. Institute Electricity Electronics Engineers, New York, 978-1-4244-9312-8/11
4. Davies JR, Coggeshall SV, Jones RD, Schutzer D (1995) Intelligent security systems. In: Freedman RS et al (eds) Artificial intelligence in the capital markets. Irwin, Chicago. ISBN 1-55738-811-3
5. Mountrakis G, Zhuang W (2012) Integrating local and global error statistics for multiscale RBF networks training. PLOS. www.plosone.org/article/info%3Adoi%2F10.1371. Accessed 21 Aug 2012

6. Balasubramanie P, Florence ML (2009) Application of radial basis network model for HIV/AIDS regimen specifications. J Comput 1:136–140
7. Kestler HA, Schwenker F, Palm G (2012) RBF network classification of ECGs as a potential marker for sudden cardiac death. www.researchgate.net/publicatrion/29528327. Accessed 21 Aug 2012
8. Hartman EJ, Keeler JD, Kowalski JM (1990) Layered neural networks with Gaussian hidden units as universal approximations. Neural Comput 2:210–215
9. Gallegos FJ, Gomez ME, Lopez JL, Cruz R (2009) Rank m-type radial basis function neural network for Pap smear microscopic image classification. Apeiron 16:542–554
10. SPSS Statistical Software. www.spss.com
11. Cleophas TF, Cleophas TJ (2011) Artificial intelligence for diagnostic purposes; principles, procedures, and limitations. Clin Chem Lab Med 48:159–165
12. Mitchell D, Strydom NB, Van Graan CH, Van der Walt H (1971) Human surface area: comparison of the du Bois formula with direct photometric measurement. Eur J Physiol 325:188–190
13. Haycock GB, Schwarz GJ, Wisotsky DH (1978) Body surface area calculated from the height and weight. J Pediatr 93:62–66
14. Atiqi R, Van Iersel C, Cleophas TJ (2009) Accuracy of quantitative diagnostic tests. Int J Clin Pharmacol Ther 47:153–159
15. Eftekbar B, Mohammad K, Ardebilli HE, Ghodsi M, Ketabchi E (2005) Comparison of artificial neural network and regression models for prediction of mortality in head trauma based on clinical data. BMC Med Inf Decis Mak 5:3–9
16. Baxt WG, Skora J (1996) Prospective validation of artificial neural network trained to identify acute myocardial infarction. Lancet 347:12–15
17. Heden B, Edenbrandt L, Hasity WK, Pahlm O (1994) Artificial neural networks for electrocardiographic diagnosis of healed myocardial infarction. Am J Cardiol 74:5–8
18. Redding NJ, Kowalczyk A, Downs T (1993) Constructive higher order network algorithms that is polynomial time. Neural Netw 6:997–1010
19. Sperduti A, Starita A (1993) Speed up learning and network optimization with extended back propagation. Neural Netw 6:365–383
20. Wnek J, Michalski RS (1994) Hypothesis driven constructive induction in AQ17-HCI: a method and experiments. Mach Learn 14:139–168
21. Lytton WW (2002) From artificial neural network to realistic neural network, Chapter 14. In: From computer to brain. Springer, New York, pp 259–268

Chapter 14
Factor Analysis

1 Summary

1.1 Background

Many factors in life are complex and difficult to measure directly. Often add-up scores of separate aspects of the complex factors are used. However, add-up scores do not account for their relative importance, their interactions and differences in units. Factor analysis accounts all of this, but is, virtually, unused in clinical research.

1.2 Objective and Methods

Using a simulated example of 200 patients at risk of septic death, we will assess whether the performance of a diagnostic battery for making clinical predictions can be improved by using factor analysis from SPSS's Data Dimension Reduction module.

1.3 Results

A three factor factor-analysis was performed. Factor 1 had a very strong correlation-ship with gammaGT, ALAT, and ASAT with correlation coefficients of 0.885, 0.878, and 0.837, the factor 2 with creatinine clearance, creatinine, and ureum (0.819, 0.694, 0.677), the factor 3 with c-reactive protein, leucos (leucocyte count), and esr (erythrocyte sedimentation rate) (0.964, 0.699, 0.658). Multiple logistic regression with death due to sepsis as outcome and the original variables (the diagnostic tests) as predictors was unable to demonstrate any significant effect.

In contrast, multiple logistic regression with the fabricated factors as predictors and again death due to sepsis as outcome demonstrated that all of the novel factors were very significant predictors (p = 0.0001, 0.001, and 0.0001).

1.4   Conclusions

Factor analysis enables to produce an improved performance of conventional diagnostic tests. It also enables to use conventional diagnostic tests for health risk profiling of individual patients. It is rather subjective, and requires biological knowledge in order to make sound decisions about clustering some of the variables and removing others.

2   Introduction

Many factors in life are complex and difficult to measure directly. Charles Spearman, a London UK psychometrician in the 1940s, searched for a method to measure intelligence [1]. Intelligence has many aspects that can be measured and modeled together. The simplest model is the use of add-up scores. However, add-up scores do not account for the relative importance of the separate aspects, their interactions and differences in units. All of this is accounted for by a technique called factor analysis: two or three unmeasured factors are identified to explain a much larger number of measured variables [2]. Although factor analysis is a major research tool in behavioral sciences, social sciences, marketing, operational research, and other applied sciences [3], it is rarely applied in clinical research. When searching the internet we found, except for a few genetic studies [4, 5], no clinical studies applying factor analysis. This is a pity given the presence of large numbers of variables in this field, particularly, in diagnostic research.

In the current chapter we will assess whether the performance of a diagnostic battery for making clinical predictions can be improved by using factor analysis.

We will also assess, whether factor analysis enables to make predictions about individuals. We hope that this chapter will stimulate clinical investigators to start using this method. For their convenience the data file of the example is given in the appendix, and each step in SPSS statistical software [2] is described.

3   Some Terminology

3.1   Internal Consistency Between the Original Variables Contributing to a Factor

There should be a strong correlation between the answers given to questions within one factor: all of the questions should, approximately, predict one and the same thing. The level of correlation is expressed as Cronbach's alpha: 0 means poor, 1 perfect relationship. The Test-retest reliability of the original variables should be assessed with one variable missing: all of the data files with one missing variable should produce at least for 80% the same result as that of the non-missing data file (alphas >80%).

3.2   Cronbach's Alpha

$$\text{alpha} = \frac{k}{(k-1)} \cdot \left(1 - \sum \frac{s_i^2}{s_T^2}\right)$$

K = number of original variables
s_i^2 = variance of ith original variable
s_T^2 = variance of total score of the factor obtained by summing up all of the original
   variables

3.3   Multicollinearity or Collinearity

There should not be a too strong correlation between different original variable values in a conventional linear regression. Correlation coefficient (R) >0.80 means the presence of multicollinearity and, thus, of a flawed multiple regression analysis.

3.4   Pearson's Correlation Coefficient (R)

$$R = \frac{\sum (x - \bar{x})(y - \bar{y})}{\sqrt{\sum (x - \bar{x})^2 \sum (y - \bar{y})^2}}$$

R is a measure for the strength of association between two variables. The stronger the association, the better one variable predicts the other. It varies between −1 and +1, zero means no correlation at all, −1 means 100% negative correlation, +1 100% positive correlation.

3.5   Factor Analysis Theory

ALAT (alanine aminotransferase), ASAT (aspartate aminotransferase) and gammaGT (gamma glutamyl tranferase) are a cluster of variables telling us something about a patient's liver function, while ureum, creatinine (creat) and creatininine clearance (c-clear) tell us something about the same patient's renal function. In order to make morbidity/mortality predictions from such variables, often, multiple regression is used. Multiple regression is appropriate, only, if the variables do not correlate too much with one another, and a very strong correlation, otherwise called collinearity or multicollinearity, will be observed within the above two clusters of variables. This means, that the variables cannot be used simultaneously in a regression model, and, that an alternative method has to be used. With factor analysis all of the variables are replaced with a limited number of novel variables, that have the largest possible correlation coefficients with all of the original variables. It is a multivariate technique, somewhat similar to MANOVA (multivariate analysis of variance), with the novel variables, otherwise called the factors, as outcomes, and the original variables, as predictors. However, it is less affected by collinearity, because the y- and x-axes are used to present the novel factors in an orthogonal way, and it can be shown that with an orthogonal relationship between two variables the magnitude of their covariance is zero and has thus not to be taken into account. Figure 14.1 shows the relationships of the six original variables with the two novel variables, the liver and renal functions. It can be observed that ureum, creatinine and creatinine clearance have a very strong positive correlation with renal function and the other three with liver function. It can be demonstrated, that, by slightly rotating both x and y-axes, the model can be fitted even better.

3.6   Magnitude of Factor Value for Individual Patients

The magnitude of the factors for individual patients are calculated as shown.

$$
\begin{aligned}
\text{Factor}_{\text{liver function}} &= 0.87 \times \text{ASAT} + 0.90 \times \text{ALAT} + 0.85 \times \text{GammaGT} + 0.10 \\
&\quad \times \text{creatinine} + 0.15 \times \text{creatininine clearance} - 0.05 \times \text{ureum} \\
\text{Factor}_{\text{renal function}} &= -0.10 \times \text{ASAT} - 0.05 \times \text{ALAT} + 0.05 \times \text{GammaGT} + 0.91 \\
&\quad \times \text{creatinine} + 0.88 \times \text{creatininine clearance} + 0.90 \times \text{ureum}
\end{aligned}
$$

3.7   Factor Loadings

The factor loadings are the correlation coefficients between the original variables and the estimated novel variable, the latent factor, adjusted for all of the original variables, and adjusted for eventual differences in units.

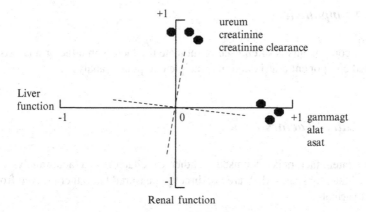

Fig. 14.1 The relationships of six original variables with two novel variables, the liver and renal function. It can be observed that ureum, creatinine and creatinine clearance have a very strong positive correlation with renal function and the other three with liver function. It can be demonstrated that by slightly rotating both x and y-axes the model can be fitted even better

3.8 Varimax Rotation

It can be demonstrated in a "2 factor" factor analysis, that, by slightly rotating both x and y-axes, the model can be fitted even better. When the y- and x-axes are rotated simultaneously, the two novel factors are assumed to be 100% independent of one another, and this rotation method is called varimax rotation. Independence needs not be true, and, if not true, the y-axis and x-axis can, alternatively, be rotated separately in order to find the best fit model for the data given.

3.9 Eigenvectors

Eigenvectors is a term often used with factor analysis. The R-values of the original variables versus novel factors are the eigenvalues of the original variables, their place in the graph (Fig. 14.1) the eigenvectors. The scree plot compares the relative importance of the novel factors, and that of the original variables using their eigenvector values.

3.10 Iterations

Complex mathematical models are often laborious, so that even modern computers have difficulty to process them. Software packages currently make use of a technique called iterations: five or more calculations are estimated and the one with the best fit is chosen.

3.11   Components

The term components is often used to indicate the factors in a factor analysis, e.g., in rotated component matrix and in principle component analysis.

3.12   Latent Factors

The term latent factors is often used to indicate the factors in a factor analysis. They are called latent, because they are not directly measured but rather derived from the original variables.

3.13   Multidimensional Modeling

An y- and x-axis are used to represent two factors. If a third factor existed within a data file, it could be represented by a third axis, a z-axis creating a 3-d graph. Also additional factors can be added to the model, but they cannot be presented in a 2- or 3-d drawing, but, just like with multiple regression modeling, the software programs have no problem with multidimensional calculations similar to the above 2-d calculations.

4   Example

A simulated example of 200 patients at risk of septic death is used (Appendix). The individual patient data are given in the appendix. We will apply SPSS statistical software [2]. We will first test the test-retest reliability of the original variables. The test-retest reliability of the original variables should be assessed with Cronbach's alphas using the correlation coefficients after deletion of one variable: all of the data files should produce at least by 80% the same result as that of the non-deleted data file (alphas >80%).

Command: Analyze....Scale....Reliability Analysis....transfer original variables to Variables box....click Statistics....mark Scale if item deleted....mark Correlations....Continue....OK.

The Table 14.1 shows, that, indeed, none of the original variables after deletion reduces the test-retest reliability. The data are reliable. We will now perform the factor analysis with three factors in the model, one for liver, one for renal and one for inflammatory function.

Table 14.1 The test-retest reliability of the original variables should be assessed with Cronbach's alphas using the correlation coefficients with one variable missing: all of the missing data files should produce at least by 80% the same result as those of the non-missing data files (alphas >80%)

	Item-total statistics				
	Scale mean if item deleted	Scale variance if item deleted	Corrected item-total correlation	Squared multiple correlation	Cronbach's alpha if item deleted
gammagt	650,6425	907820,874	,892	,827	,805
asat	656,7425	946298,638	,866	,772	,807
alat	660,5975	995863,808	,867	,826	,803
bili	789,0475	1406028,628	,938	,907	,828
ureum	835,8850	1582995,449	,861	,886	,855
creatinine	658,3275	1244658,421	,810	,833	,814
creatinine clearance	929,7875	1542615,450	,721	,688	,849
esr	812,2175	1549217,863	,747	,873	,850
c-reactive protein	827,8975	1590791,528	,365	,648	,857
leucos	839,0925	1610568,976	,709	,872	,859

Command: Analyze….Dimension Reduction….Factor….enter variables into Variables box….click Extraction….Method: click Principle Components…. mark Correlation Matrix, Unrotated factor solution….Fixed number of factors: enter 3….Maximal Iterations for Convergence: enter 25….Continue….click Rotation….Method: click Varimax….mark Rotated solution….mark Loading Plots….Maximal Iterations: enter 25….Continue….click Scores…. mark Display factor score coefficient matrix ….OK.

The Table 14.2 shows the best fit coefficients of the original variables constituting three components (otherwise called factors here), that can be interpreted as overall liver function (1), renal function (2), and inflammatory function (3) as calculated by the factor analysis program. The component 1 has a very strong correlation with gammaGT, ALAT, and ASAT, the component 2 with creatinine clearance, creatinine, and ureum, the component 3 with c-reactive protein, leucos (leucocyte count), and esr (erythrocyte sedimentation rate).

The three components can, thus, be interpreted as overall liver function (1), renal function (2), and inflammatory function (3). When minimizing the outcome file and returning to the data file, we now observe, that, for each patient, the software program has produced the individual values of the factors 1 (liver function), 2 (renal function), and 3 (inflammatory function).

When performing multiple logistic regression with death due to sepsis as outcome variable and the original variables (the diagnostic tests) as predictor, we are unable to demonstrate a significant effect of any of the variables on death due to sepsis (Table 14.3). In contrast, using the fabricated factors as predictors the effects were

Table 14.2 The coefficients of the variables that constitute three components that can be interpreted as overall liver function (1), renal function (2), and inflammatory function (3) as calculated by the factor analysis program

	Rotated component matrix[a]		
	Component		
	1	2	3
gammagt	,885	,297	,177
alat	,878	,297	,167
bili	,837	,423	,215
asat	,827	,339	,206
creatinine clearance	,373	,819	,204
creatinine	,555	,694	,262
ureum	,590	,677	,285
c-reactive protein	,105	,102	,964
leucos	,325	,572	,699
esr	,411	,546	,658

Extraction method: Principal component analysis
Rotation method: Varimax with Kaiser normalization
Gammagt gamma glutamyl tranferase (N < 50 U/l), *alat* alanine aminotransferase (N < 41 U/l), *bili* total bilirubine (N < 5 mmumol/l), *asat*
aspartate aminotransferase (N < 37 U/l), *leucos* leucocyte count
(N < 10.10^9/l), *esr* erythrocyte sedimentation rate (N < 20 mm)
[a]Rotation converged in five iterations

Table 14.3 Binary logistic regression with the traditional clinical chemistry tests as independent and the odds of death from sepsis as dependent variable (VAR0001)

		B	S.E.	Wald	df	Sig.	Exp(B)
		Variables in the equation					
Step 1[a]	VAR00002	,266	8,384	,001	1	,975	1,304
	VAR00003	−,063	3,193	,000	1	,984	,939
	VAR00004	−,054	10,128	,000	1	,996	,947
	VAR00005	,733	43,310	,000	1	,987	2,081
	VAR00006	3,850	89,435	,002	1	,966	46,991
	VAR00007	−,423	36,954	,000	1	,991	,655
	VAR00008	−,518	56,161	,000	1	,993	,596
	VAR00009	−,038	97,714	,000	1	1,000	,963
	VAR00010	21,716	333,738	,004	1	,948	2,699E9
	VAR00011	1,561	333,798	,000	1	,996	4,763
	Constant	−379,414	8219,991	,002	1	,963	,000

VAR = variable; var 1 = death (1 = yes); var 2 = gammagt; var 3 = alat; var 4 = asat; var 5 = bili; var 6 = ureum; var 7 = creatinine; var 8 = creatinine clearance; var 9 = erythrocyte sedimentation rate; var 10 = c-reactive protein; var 11 = leucos
[a]Variable(s) entered on step 1: VAR00002, VAR00003, VAR00004, VAR00005, VAR00006, VAR00007, VAR00008, VAR00009, VAR00010, VAR00011

very significant. Table 14.4 shows the results: logistic regression with the factors as predictors and again death due to sepsis as outcome variable demonstrates that all of the novel factors are very significant independent predictors of the risk of death.

Table 14.4 Binary logistic regression with the factors produced by the factor analysis model as independent and the odds of death as dependent variable

		Variables in the equation				
		B	S.E.	Wald	df	Sig.
Step 1[a]	FAC1_1	3,032	,627	23,393	1	,000
	FAC2_1	1,900	,560	11,503	1	,001
	FAC3_1	8,933	1,713	27,196	1	,000
	Constant	2,597	,623	17,391	1	,000

[a]Variable(s) entered on step 1: FAC1_1, FAC2_1, FAC3_1

5   Making Predictions for Individual Patients, Health Risk Profiling

The individual values of the factors 1, 2, and 3 can now also be used to assess the risk of morbidity/mortality for individual patients. As an example: the data of a patient with the underneath characteristics can be added at the bottom of the file.

Ggt	asat	alat	bili	ureum	creatinine	c-clear	esr	crp	leucos
257,00	149,00	123,00	73,00	16,00	154,00	−45,00	36,00	25,00	16,00

Command: Analyze….Dimension Reduction…Factor Analysis…. OK.

The following factor values are displayed for this patient.

Factor 1	Factor 2	Factor 3
−0.33216	0.59800	−0,00371

$$\text{Log odds death} = 3.03 \times (-0.33) + 1.90 \times 0.60 - 8.93 \times 0.004$$
$$= -0.999 + 1.140 - 0.036$$
$$= 0.105$$
$$\text{odds} = \text{antilog } 0.111$$
$$= 1.19$$
$$\text{risk} = \text{odds} / (1 + \text{odds})$$
$$= 53\,\%.$$

The risk of death for a patient with these variables will be 53%. Similarly, the risk of death can be calculated from the underneath patients.

Ggt	asat	alat	bili	ureum	creatinine	c-clear	esr	crp	leucos
276,00	200,00	154,00	72,00	15,00	243,00	−55,00	38,00	24,00	16,00
276,00	230,00	156,00	79,00	17,00	235,00	−54,00	34,00	23,00	15,00
269,00	264,00	267,00	69,00	19,00	244,00	−55,00	45,00	29,00	15,00

Their risks of death can be calculated to be respectively 59, 43, and 85%. Log linear risk profiling is, currently, an important method to determine, with limited health care sources, what patients will be given expensive treatments or not, as well as other treatment decisions.

6   Discussion

Factor analysis in clinical trials serves two purposes. It enables to produce an improved performance of conventional diagnostic tests. It enables to use the conventional diagnostic tests for health risk profiling of individual patients. Except for sporadic use in genetic research, it is, virtually, unused in clinical research. This is a pity, since factor analysis has been demonstrated to be easy and inexpensive to do, and, because it increases the statistical power of testing by reducing a multiple variables model into a sparse variables model, thereby reducing the risk of false positive effects. Another advantage of factor analysis is, that it can, sometimes, discover hidden interrelationships between variables.

We should add some limitations.

1. The original variables in a factor should be strong-positive correlated, and this issue should be tested prior to the analysis, e.g. by using Cronbach's alphas of the complete data versus the data after one by one deletion of the original variables.
2. Background knowledge or a theory about the mechanisms causing the strong correlations is required; high correlation without an apparent reason or without an adequate underlying theory, is, clinically, generally not very relevant, and could easily be due to type I errors.
3. Correlationships should not be taken equal for causalities. In order to prove causality the introduction of a factor and the occurrence of a subsequent event is considered better proof for that purpose.

The example used in this chapter must be confirmed with real data. Often the correlations between the clusters of original variables are smaller than in the example given. It is, then, difficult to choose which variables to maintain in the model and which to remove, also which variables to add to what factor. Like with other aspects of statistics choices to be made are rather subjective. Statistics requires a lot of biological knowledge and a bit of math used to answer biological questions. Particularly factor analysis is a good example of this principle.

7   Conclusions

1. To date factor analysis is, virtually, unused in clinical research.
2. Factor analysis enables to produce an improved performance of conventional diagnostic tests.
3. It also enables to use the conventional diagnostic tests for health risk profiling of individual patients.
4. It is rather subjective and requires biological knowledge in order to make sound decisions about clustering some of the variables and removing others.

Appendix: Data File of 200 Patients Admitted Because of Sepsis

Death	Ggt	asat	alat	bili	ureum	creat	c-clear	esr	crp	leucos
var 1	var 2	var 3	var 4	var 5	var 6	var 7	var 8	var 9	var 10	var 11
0	20	23	34	2	3,4	89	−111	2	2	5
0	14	21	33	3	2	67	−112	7	3	6
0	30	35	32	4	5,6	58	−116	8	4	4
0	35	34	40	4	6	76	−110	6	5	7
0	23	33	22	4	6,1	95	−120	9	6	6
0	26	31	24	3	5,4	78	−132	8	4	8
0	15	29	26	2	5,3	47	−120	12	5	5
0	13	26	24	1	6,3	65	−132	13	6	6
0	26	27	27	4	6	97	−112	14	6	7
0	34	25	13	3	4	67	−125	15	7	6
0	32	26	24	3	3,6	58	−110	13	8	6
0	21	13	15	3	3,6	69	−102	12	2	4
0	16	14	12	3	3,2	87	−124	11	3	3
0	17	15	8	2	4,5	50	−112	9	4	2
0	19	16	9	4	4,2	80	−113	8	4	7
0	16	17	10	3	4,9	90	−112	18	4	8
0	26	26	34	1	4,7	65	−117	17	3	9
0	25	25	25	3	5,7	76	−107	17	5	6
0	24	24	27	4	4,2	84	−120	15	6	6
0	30	23	26	4	5,8	67	−120	18	4	5
0	18	12	25	3	5,3	68	−101	15	7	4
0	2	13	24	3	4,8	56	−108	16	6	7
0	9	14	15	3	4,2	76	−111	17	5	8
0	9	17	14	2	5,3	96	−112	12	8	6
0	5	16	13	2	3,2	99	−116	13	7	4
0	7	15	12	2	4,2	89	−110	14	6	7
0	2	18	38	3	5,4	70	120	19	9	6
0	20	16	37	3	5,9	101	−132	18	4	6
0	13	13	36	3	5,8	65	−120	8	5	8
0	14	3	35	3	4,7	87	−132	9	7	9
0	16	5	24	4	4,6	69	−112	7	3	4
0	14	6	25	3	4,3	88	−125	8	4	3
0	12	4	26	4	3,2	80	−110	9	4	2
0	14	3	27	2	3	97	−102	7	5	5
0	16	8	13	3	5	79	−124	8	7	5
0	20	23	14	2	3,6	86	−112	9	7	6
0	19	236	15	2	3,4	78	−113	7	6	6
0	27	25	16	2	3,5	845	−112	12	8	7
0	16	24	17	3	3,8	83	−117	14	8	4
0	26	3	24	3	4,2	94	−107	13	7	6
0	25	24	25	3	4,3	82	−120	15	7	6
0	24	12	26	4	5,7	83	−120	15	7	4

(continued)

(continued)

Death	Ggt	asat	alat	bili	ureum	creat	c-clear	esr	crp	leucos
var 1	var 2	var 3	var 4	var 5	var 6	var 7	var 8	var 9	var 10	var 11
0	27	17	27	4	5,3	98	−101	14	4	3
0	23	15	28	4	5,4	83	−108	16	5	2
0	20	13	14	3	5,8	84	−132	13	6	7
0	17	13	13	3	5,4	88	−120	17	7	8
0	16	12	12	3	5,3	82	−132	14	6	9
0	15	17	15	2	6,4	89	−112	13	9	6
0	15	15	25	2	3,4	72	−125	15	4	6
0	14	14	24	2	2	78	−110	12	5	5
0	13	13	23	1	5,6	73	−102	16	7	4
0	18	19	19	1	6	79	−124	7	3	7
0	26	28	18	1	6,1	74	−112	8	4	8
0	37	17	17	3	5,4	78	−113	7	4	6
0	25	26	16	4	5,3	79	−112	16	5	4
0	36	15	29	2	6,3	89	−117	15	7	7
0	35	27	8	3	6	67	−107	17	7	6
0	24	14	27	4	4	76	−120	8	6	6
0	15	22	26	2	3,6	75	−120	9	8	8
0	17	21	16	4	3,6	87	−101	9	8	9
0	28	11	17	3	3,2	77	−108	13	7	4
0	29	13	18	2	4,5	88	−100	15	7	3
0	16	14	19	4	4,2	67	−102	14	7	2
0	18	16	26	3	4,9	87	−105	17	4	5
0	14	14	27	2	4,7	76	−109	18	5	5
0	15	18	28	4	5,7	87	−108	16	6	6
0	16	27	29	3	4,2	77	−102	14	4	6
0	24	25	24	2	5,8	69	−110	16	5	7
0	21	29	25	4	5,3	78	−112	15	7	4
0	21	24	23	3	4,8	69	−113	14	3	6
0	23	23	26	2	4,2	71	−111	14	4	6
0	24	14	25	2	5,3	72	−111	16	4	8
0	25	17	27	3	3,2	83	−113	13	5	9
0	21	15	13	2	4,2	92	−120	17	7	4
0	13	14	15	3	5,4	93	−119	14	7	3
0	14	18	4	2	5,9	91	−117	13	6	2
0	15	25	3	2	5,9	82	−114	15	8	5
0	16	26	2	2	5,8	83	−120	12	8	5
0	25	23	15	2	4,7	79	−131	16	7	6
0	34	24	14	3	4,6	89	−120	7	7	6
0	23	23	26	3	4,3	67	−102	8	7	7
0	15	23	14	3	3,2	76	−105	7	4	4
0	36	15	27	3	3	75	−109	16	5	4
0	24	14	14	4	5	87	−108	15	6	7
0	34	27	27	4	3,6	77	−102	17	7	6
0	21	27	15	3	3,4	88	−110	8	6	8
0	35	25	14	4	3,5	67	−112	9	8	5
0	23	25	24	2	3,8	87	−113	9	8	6

(continued)

(continued)

Death	Ggt	asat	alat	bili	ureum	creat	c-clear	esr	crp	leucos
var 1	var 2	var 3	var 4	var 5	var 6	var 7	var 8	var 9	var 10	var 11
0	23	14	15	3	4,2	76	−111	13	7	7
0	25	26	26	4	4,3	87	−111	15	7	6
0	24	25	15	2	5,7	77	−113	14	7	6
0	15	14	26	3	5,3	69	−120	17	4	4
0	26	25	14	4	5,4	78	−119	18	5	3
0	26	34	27	2	5,8	69	−117	16	6	2
0	27	33	15	3	4,3	71	−114	14	4	7
0	14	30	24	4	4,1	72	−120	16	5	8
0	25	29	15	3	3,8	83	−131	15	7	9
0	14	28	26	2	3,8	92	−120	14	3	6
0	13	18	23	1	3,5	93	−110	12	4	6
0	25	16	29	201	4,8	91	−102	14	4	5
1	609	599	1,500	304	3,5	93	−110	12	4	6
1	1,500	709	1,154	400	46	601	−10	7	7	8
1	1,453	876	1,342	254	58	606	−12	150	7	25
1	25	16	29	201	90	645	−13	180	6	24
1	757	588	1,453	344	98	765	−14	143	8	30
1	566	690	1,432	233	76	625	−14	112	8	30
1	899	588	1,276	204	3,8	92	−120	14	3	6
1	2,300	800	876	301	46	654	−6	109	7	23
1	1,320	1,980	845	382	76	567	−7	123	7	26
1	1,342	1,870	756	203	45	578	−4	143	4	22
1	709	1,980	645	254	42	865	−4	125	5	26
1	687	800	543	244	43	577	−6	143	6	23
1	25	29	15	3	45	566	−7	123	100	29
1	599	868	856	265	46	544	−9	165	120	30
1	900	759	856	287	45	532	−8	109	103	23
1	876	760	876	300	48	657	−5	102	90	24
1	945	879	946	266	87	685	−5	104	89	25
1	603	576	973	203	67	768	−5	104	79	24
1	1,065	1,487	865	265	63	675	−10	105	110	30
1	26	34	27	2	54	666	−10	108	124	30
1	1,243	577	675	253	65	765	−11	116	87	29
1	1,258	900	774	287	80	845	−12	157	84	23
1	1,600	798	766	376	65	503	−7	120	83	26
1	25	17	27	3	3,2	83	−113	123	140	30
1	1,208	1,500	958	387	69	534	−14	119	187	26
1	340	450	301	165	32	300	−20	61	34	18
1	498	436	344	176	33	345	−30	65	35	18
1	345	465	365	134	24	450	−34	65	36	19
1	376	388	465	135	23	342	−20	69	45	19
1	354	366	499	186	25	260	−23	83	46	19
1	387	457	354	124	34	276	−26	78	47	20
1	399	497	365	125	35	265	−25	76	47	18
1	25	16	29	201	38	345	−22	56	46	22
1	454	398	477	124	35	376	−21	65	45	21

(continued)

(continued)

Death	Ggt	asat	alat	bili	ureum	creat	c-clear	esr	crp	leucos
var 1	var 2	var 3	var 4	var 5	var 6	var 7	var 8	var 9	var 10	var 11
1	497	355	476	136	25	333	27	45	36	20
1	477	389	365	154	24	334	−32	54	85	22
1	459	450	345	143	28	423	−30	34	34	21
1	432	487	376	175	26	260	−32	43	32	21
1	412	499	498	164	23	287	−31	76	34	21
1	302	378	499	145	26	299	−23	87	43	19
1	306	320	376	156	25	246	−15	98	45	19
1	376	421	465	176	36	256	−18	67	43	18
1	385	435	387	145	35	276	−23	86	34	17
1	320	428	465	165	34	299	−24	78	34	19
1	376	459	389	135	25	267	−29	97	33	20
1	345	490	356	176	26	256	−23	66	34	18
1	14	3	35	3	4,7	87	−132	120	135	30
1	24	25	15	2	29	599	−16	75	36	21
1	14	28	26	2	3,8	92	−120	74	137	29
1	28	11	17	3	3,2	77	−108	132	185	30
1	159	180	287	56	14	240	−40	40	20	14
1	169	154	267	75	13	244	−50	42	21	15
1	178	176	265	46	19	235	−42	43	23	16
1	180	287	286	85	18	224	−43	50	29	16
1	194	277	257	67	17	231	−48	59	28	16
1	175	250	276	95	17	231	−41	36	28	15
1	13	14	15	3	18	241	−42	45	25	17
1	21	27	15	3	3,4	88	−110	160	243	29
1	164	199	177	66	14	176	−38	46	27	18
1	174	256	186	67	18	165	−39	38	25	18
1	188	234	156	63	16	174	−42	45	28	17
1	196	285	197	65	15	165	−59	46	24	16
1	275	237	167	84	19	177	−56	43	28	16
1	266	263	187	54	14	198	−45	42	25	16
1	245	296	287	60	12	154	−54	48	24	15
1	290	188	165	67	14	156	−53	47	26	14
1	275	182	167	69	14	187	−57	45	27	15
1	267	139	177	74	13	167	−48	37	28	14
1	257	149	123	73	16	154	−45	36	25	16
1	276	200	154	72	15	243	−55	38	24	16
1	276	230	156	79	17	235	−54	34	23	15
1	269	264	267	69	19	244	−55	45	29	15
1	162	276	287	87	16	175	−36	46	26	16
1	263	287	234	85	16	165	−37	47	27	16
1	172	288	265	83	13	149	−38	34	25	15
0	100	80	47	14	9	134	−61	29	12	13
0	102	53	67	17	8	144	−70	28	14	12
1	80	63	87	45	7	132	−65	28	15	11
0	89	72	120	43	9	127	−85	28	13	12
1	120	82	143	44	8	136	−80	,26	16	12

(continued)

(continued)

Death	Ggt	asat	alat	bili	ureum	creat	c-clear	esr	crp	leucos
var 1	var 2	var 3	var 4	var 5	var 6	var 7	var 8	var 9	var 10	var 11
1	143	87	67	37	7,4	128	−70	20	17	14
1	75	84	145	39	7,6	137	−66	28	18	14
1	79	73	149	47	7,3	133	−68	27	16	13
1	99	93	100	49	8,7	137	−66	26	16	14
0	92	110	108	27	8,9	142	−72	29	14	13
0	79	129	109	26	9	136	−73	23	13	12
1	130	143	105	25	9,4	142	−70	24	15	11
1	134	145	90	39	7,6	149	−67	25	18	12
1	149	137	93	38	9	143	−80	23	19	12
1	134	142	95	48	8	138	−67	24	16	14
1	145	130	108	46	7	132	−65	25	16	14
1	123	90	102	43	9	136	−76	23	17	13
1	27	33	15	3	4,3	71	−114	24	19	14
1	95	56	105	41	7,4	148	−65	25	15	13
0	134	58	109	46	7,6	143	−64	27	13	12
1	125	49	93	48	7,3	136	−64	25	16	11
1	23	25	24	2	3,8	87	−113	28	16	12
1	129	76	95	46	8,9	144	−68	26	18	12
1	25	23	15	2	4,7	79	−131	29	19	14
1	16	5	24	4	4,9	90	−112	27	15	14

1 = yes

References

1. Barthelemew DJ (1995) Spearman and the origin and development of factor analysis. Br J Math Stat Psychol 48:211–220
2. Anonymous (2012) SPSS statistical software version 18.0. Module dimension reduction, factor analysis, Online Help. www.spss.com. Accessed 29 Apr 2012
3. Anonymous (2012) Factor analysis. Wikipedia, the free encyclopedia, Accessed 25 Apr 2012
4. Meng J (2012) Uncover cooperative gene regulations by microRNAs and transcription factors in glioblastoma using a nonnegative hybrid factor. www.cmsworldwide.com?ICASS2011. Accessed 29 Apr 2012
5. Hochreiter S, Clevert DA, Obermayer K (2006) A new summarization method for affymetrix probe level data. Bioinformatics 22:943–949

Chapter 15
Hierarchical Cluster Analysis for Unsupervised Data

1 Summary

1.1 Background

Drug efficacy is multifactorial and with multiple variables regression modeling rapidly looses power and it is invalid if the correlations between the variables is strong. Hierarchical cluster analysis can handle hundreds of variables, and is unaffected by strong correlations.

1.2 Objective

To assess its performance.

1.3 Methods

A simulated example of 11 unrelated gastric cancer patients (here called cases) sensitive to cytostatic treatment, with nine highly expressed genes was used. For analysis SPSS statistical software was used.

1.4 Results

Two clusters of cases were indentified with cases more similar to one another than to the cases of the other cluster. In a linear regression cluster membership was a

T.J. Cleophas and A.H. Zwinderman, *Machine Learning in Medicine*,
DOI 10.1007/978-94-007-5824-7_15, © Springer Science+Business Media Dordrecht 2013

predictor of progression free interval at $p < 0.0001$. When comparing these results with those of factor analysis, another machine learning method, the result was somewhat similar although at a lower level of significance with p-values of 0,002 and 0,143.

1.5   Conclusions

1. Drug efficacy is not a simple monocausal entity, but rather a complex one of a multifactorial nature.
2. Traditional clinical trials can only count a mean result versus control, and are unable to assess multiple factors.
3. Hierarchical cluster analysis is able to do so, and performed well in the example given: two clusters of patients were identified with significantly different progression free intervals.
4. Hierarchical cluster analysis may be more appropriate for drug efficacy analyses than other machine learning methods, like factor analysis, because the patients themselves rather than some new characteristics are used as dependent variables.

2   Introduction to Hierarchical Cluster Analysis

2.1   A Novel Approach to Drug Efficacy Testing

Traditionally, drug efficacy is tested in clinical trials of relatively homogenous groups. However, this approach can only count a mean result versus control, and is unable to assess multiple predictor variables. An entirely different approach is to assess populations with high drug efficacy for their particular characteristics. This approach is relevant considering the multifactorial nature of drug efficacy [1–5]. E.g., patients with high efficacy of cytostatic treatments had high expression levels of a number of genes [2, 3], patients with high HIV vaccine efficacy had particular clinical characteristics [4], and in patients with high anti-trypanosomal drug efficacy several underlying mechanisms were established [5]. Many more examples can be given. In this kind of research a population is primarily selected based on its treatment results, and the investigators have some idea about particular patient characteristics responsible for their high drug efficacy, for example a selected set of gene expressions, physical symptoms, other personal characteristics like gender, age, co-morbidities and co-medications etc. What is the clinical relevance of this kind of research. Subgroups may be established with different prognoses, treatment susceptibilities or resistance, time to progression etc.

2.2 A Novel Statistical Methodology Suitable for the Purpose

Commonly, linear regression is applied for analysis, but with multiple characteristics this method rapidly looses power, and it is not valid with strong correlations between the characteristics, otherwise called collinearity. Hierarchical cluster analysis can handle hundreds of variables, and is unaffected by collinearity, because it is not based on correlations but on distances between individual patients. It was invented by Robert Sibson, statistician from King's College Cambridge UK statistical department [6] and Daniel Defays, psychologist from Liege University Belgium [7], and was initially used by sociologists and psychologists for identifying similarities in population subgroups [8].

2.3 Publications So Far

Apart from its current key role in sequence-clustering [9, 10], which is a method for clustering related DNA and protein sequences, we found only sporadic published papers of cluster analysis in treatment efficacy studies. Two adverse event studies [11, 12], have been published, respectively one study of the effects of subsets of genes on chemo-resistance to platinum and fluorouracil, and one general study of frequently observed drug adverse effects. One drug manufacturing study of the effects of 57 pharmaceutical solvents on crystallization optimization [13], and one patient compliance study of adherence factors to oral thiopurines have been published [14].

2.4 Objective of the Current Chapter

In the current chapter, using a simulated case-study of highly expressed genes on cytostatic drug efficacy, we will assess the performance of hierarchical clustering for drug efficacy analysis. The chapter was written as a hand-hold presentation accessible to clinicians, and as a must-read paper for those new to the method. It is my experience as masters' class teacher that students are eager to master adequate command of statistical software. For their benefit all of the steps from logging in to the final result using SPSS statistical software [15] are given.

3 Case-Study

3.1 Example

In 11 unrelated gastric cancer patients (here called cases) sensitive to cytostatic treatment, nine highly expressed genes (here called genes 13–21) were sampled. The levels of gene expression were scored on 0–10 linear scales (Table 15.1). A correlation matrix of the variables was constructed using SPSS.

Table 15.1 Data file: the expression levels of 9 selected genes (scores 0,00–10,00) and time to progression (0 = short, 1 = long) in 11 unrelated gastric cancer patients (cases) with high cytostatic drug efficacy

gene 13	gene 14	gene 15	gene 16	gene 17	gene 18	gene 19	gene 20	gene 21	
									Progression free interval (months)
4,42	4,40	5,75	5,21	6,79	9,05	2,68	7,15	6,75	15
5,51	6,20	8,50	5,35	6,94	9,52	3,05	7,50	6,25	12
6,30	6,80	9,00	5,53	7,30	10,00	3,39	8,50	6,75	8
7,19	7,60	9,50	5,06	7,31	9,16	3,20	7,85	6,00	9
5,96	6,00	7,75	5,43	7,30	9,88	3,37	8,00	6,00	10
4,10	4,00	5,35	5,15	6,69	8,74	2,56	6,60	7,50	14
4,30	3,20	5,30	5,16	6,71	8,75	2,34	5,95	8,00	16
5,12	4,60	6,75	5,35	7,03	9,44	2,93	8,55	6,75	14
6,57	6,80	8,75	5,35	7,04	9,32	3,09	7,60	6,25	12
4,37	3,60	6,00	4,85	6,67	8,70	2,42	6,60	8,25	15
5,84	4,40	6,65	5,26	7,01	9,43	3,00	9,25	6,75	16

Command: Analyze….Correlation….Bivariate….enter variables….OK.

The correlation matrix showed correlation coefficients mostly >0,85 indicating that the genes were very much correlated with one another (Table 15.2). All of the genes were oncogenic, and, we expected that patients with more similar levels of gene expressions might be more similar with regard to their progression free interval. We, first, assessed whether such subgroups could be identified. Subsequently, we tested the subgroup membership against the progression free intervals (months, data are given in Table 15.1).

3.2 Data Analysis

Patients are called cases. A cluster is estimated by the maximum distances between the values needed to connect the cases. The smaller the distance, the more similar the cases are. The distance is calculated as demonstrated in Table 15.3, by the add-up sum of squared difference between two cases. We will use SPSS statistical software [15].

Command: Analyze….Classify….Hierarchical Clustering…enter variables…. Label Case by: case variable with the values 1–11…Statistics: mark Proximity Matrix….Plots: mark Dendrogram….Method….Cluster Method: Between-group linkage….Measure: Squared Euclidean Distance….OK.

An overview of all distances between two cases is given in the proximity matrix (Table 15.4). The smallest distances are used to form the first clusters namely the

Table 15.2 The correlation matrix of the variables from Fig. 15.1 showed correlation coefficients mostly >085 indicating that the genes (variables 13–21) were very much correlated with one another

Correlations

		var 13	var 14	var 15	var 16	var 17	var 18	var 19	var 20	var 21
var 13	Pearson correlation	1	,912**	,930**	,414	,892**	,642*	,864**	,637*	-,789**
	Sig. (2-tailed)		,000	,000	,206	,000	,033	,001	,035	,004
	N	11	11	11	11	11	11	11	11	11
var 14	Pearson correlation	,912**	1	,973**	,471	,842**	,645*	,869**	,469	-,834**
	Sig. (2-tailed)	,000		,000	,144	,001	,032	,001	,146	,001
	N	11	11	11	11	11	11	11	11	11
var 15	Pearson correlation	,930**	,973**	1	,450	,838**	,671*	,859**	,513	-,767**
	Sig. (2-tailed)	,000	,000		,164	,001	,024	,001	,106	,006
	N	11	11	11	11	11	11	11	11	11
var 16	Pearson correlation	,414	,471	,450	1	,597	,862**	,713*	,566	-,608*
	Sig. (2-tailed)	,206	,144	,164		,053	,001	,014	,069	,047
	N	11	11	11	11	11	11	11	11	11
var 17	Pearson correlation	,892**	,842**	,838**	,597	1	,833**	,951**	,715*	-,799**
	Sig. (2-tailed)	,000	,001	,001	,053		,001	,000	,013	,003
	N	11	11	11	11	11	11	11	11	11
var 18	Pearson correlation	,642*	,645*	,671*	,862**	,833**	1	,911**	,766**	-,712*
	Sig. (2-tailed)	,033	,032	,024	,001	,001		,000	,006	,014
	N	11	11	11	11	11	11	11	11	11
var 19	Pearson correlation	,864**	,869**	,859**	,713*	,951**	,911**	1	,763**	-,870**
	Sig. (2-tailed)	,001	,001	,001	,014	,000	,000		,006	,001
	N	11	11	11	11	11	11	11	11	11

(continued)

Table 15.2 (continued)

Correlations

		var 13	var 14	var 15	var 16	var 17	var 18	var 19	var 20	var 21
var 20	Pearson correlation	,637*	,469	,513	,566	,715*	,766**	,763**	1	-,630*
	Sig. (2-tailed)	,035	,146	,106	,069	,013	,006	,006		,038
	N	11	11	11	11	11	11	11	11	11
var 21	Pearson correlation	-,789**	-,834**	-,767**	-,608*	-,799**	-,712*	-,870**	-,630*	1
	Sig. (2-tailed)	,004	,001	,006	,047	,003	,014	,001	,038	
	N	11	11	11	11	11	11	11	11	11

*Correlation is significant at the 0.05 level (2-tailed); **Correlation is significant at the 0.01 level (2-tailed)

Table 15.3 Example of add-up sum of squared difference between two cases (cases 2 and 5)

gene 13	gene 14	gene 15	gene 16	gene 17	gene 18	gene 19	gene 20	gene 21
5,51	6,20	8,50	5,35	6,94	9,52	3,05	7,50	6,25
5,96	6,00	7,75	5,43	7,30	9,88	3,37	8,00	6,00
Difference								
0,45	0,20	0,75	0,08	0,36	0,36	0,32	0,50	0,25
Squares								
0,20	0,04	0,56	0,006	0,13	0,13	0,10	0,25	0,06
Add-up sum of squares								
1,48								

cases [8 and 11], [7 and 10], [6 and 10], [3 and 9], [6 and 7], [2 and 5] all have distances <2,0. The cases no 6, 7, and 10 also have rather small distances between one another, namely between 1,248 and 1,402, and can, therefore, be included in a cluster of three patients. For calculation the overall distance is the average of the separate distances.

An icicle plot (named after the picture of icicles hanging from eaves) is in Fig. 15.1. It shows each step of the cluster analysis, when average linkage is used to link the clusters. The graph should be read from the bottom up. We start with 11 clusters, i.e. each patient is as cluster of its own. Then with 10 clusters case 8 and 10 have been clustered. With 9 clusters case 7 and 10, with 8 clusters case 6 and 7, with 2 clusters 8 and 10 have been clustered and only 2 and 11 are unclustered. Finally, with 1 cluster all of the cases have been clustered in a single large cluster of all cases. It can be observed considering the 2 cluster situation that the cases 2, 3, 4, 5, 9, and the cases 1, 6, 7, 8, 10, 11 are the constituents of the two clusters. Figure 15.2 shows a dendrogram from the data of Fig. 15.1. The observed distances are rescaled to fall into a range of 0–25. The cases 2, 3, 4, 5, 9, are clustered together at a rescaled distance of approximately 5, while the cases 1, 6, 7, 8, 10, 11 are clustered at approximately 12. Obviously, Two clusters of cases can be indentified with cases more similar to one another than to the cases of the other cluster.

When the two clusters were used in a linear regression as predictors of progression free interval (Table 15.5), a very significant effect was observed. The cluster with the smaller distances between cases was more at risk of short time to progression at $p < 0.0001$.

4 Discussion

4.1 Multifactorial Nature of Drug Efficacy

Drug efficacy research involves multiple variables and it is time that the simple construct of a clinical trial to predict drug efficacy from a homogeneous group was replaced or at least complemented with methodologies that enable to assess multiple

Table 15.4 The add-up sum of the squared distances between two patients (cases) is computed for all nine genes

Proximity matrix

Squared Euclidean Distance

Case	1:1	2:2	3:3	4:4	5:5	6:6	7:7	8:8	9:9	10:10	11:11
1:1	,000	12,777	23,450	33,590	11,706	1,403	4,867	3,784	20,178	3,583	7,534
2:2	12,777	,000	2,981	6,321	1,483	20,062	27,366	7,160	1,613	20,518	10,107
3:3	23,450	2,981	,000	3,619	3,152	32,919	41,856	11,925	1,855	32,296	12,692
4:4	33,590	6,321	3,619	,000	7,821	44,468	54,251	22,209	1,894	44,093	22,941
5:5	11,706	1,483	3,152	7,821	,000	19,811	27,566	5,002	2,706	21,399	6,365
6:6	1,403	20,062	32,919	44,468	19,811	,000	1,402	8,488	28,826	1,328	13,215
7:7	4,867	27,366	41,856	54,251	27,566	1,402	,000	14,018	36,862	1,248	19,083
8:8	3,784	7,160	11,925	22,209	5,002	8,488	14,018	,000	12,155	9,358	1,074
9:9	20,178	1,613	1,855	1,894	2,706	28,826	36,862	12,155	,000	28,875	13,711
10:10	3,583	20,518	32,296	44,093	21,399	1,328	1,248	9,358	28,875	,000	13,639
11:11	7,534	10,107	12,692	22,941	6,365	13,215	19,083	1,074	13,711	13,639	,000

The values estimate the distance between two patients. The smallest distances are used to form the first clusters namely patients 8 and 11, 7 and 10, 6 and 10, 3 and 9, 6 and 7, 2 and 5 all have distance <2,0. Patients no 6, 7, and 10 have rather similar small distances between 1,248 and 1,402, and can be included a cluster of three patients. For that purpose the distances are averaged

This is a dissimilarity matrix

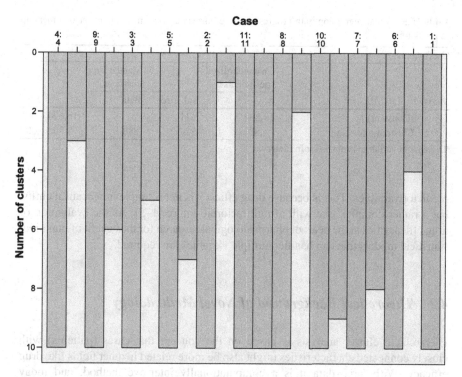

Fig. 15.1 Icicle plot (named after the picture of icicles hanging from eaves). It shows each step of the cluster analysis when average linkage is used to link the clusters. The graph should be read from the bottom up. We start with 11 clusters, i.e. each patient is as cluster of its own. Then with 10 clusters case 8 and 10 have been clustered. With 9 clusters case 7 and 10, with 8 clusters case 6 and 7, with 2 clusters 8 and 10 have been clustered and only 2 and 11 are unclustered. Finally, with 1 cluster all of the cases have been clustered in a single large cluster of all cases. It can be observed considering the 2 cluster situation that the cases 2, 3, 4, 5, 9, and the cases 1, 6, 7, 8, 10, 11 are the constituents of the two clusters

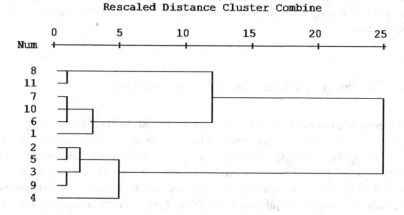

Fig. 15.2 Dendrogram from the data of Fig. 15.1. The observed distances are rescaled to fall into a range of 0–25. The cases 2, 3, 4, 5, 9, are clustered together at a rescaled distance of approximately 5, while the cases 1, 6, 7, 8, 10, 11 are clustered at approximately 12

Table 15.5 Linear regression with progression free interval as outcome and cluster membership as predictor

Coefficients[a]

Model		Unstandardized coefficients		Standardized coefficients		
		B	Std. error	Beta	t	Sig.
1	(Constant)	10,200	,611		16,694	,000
	Clustermembership	4,800	,827	,888	5,802	,000

[a]Dependent variable: progressionfreeinterval

predictor variables. This is because drug efficacy is not a simple monocausal entity, but a rather complex one with a multifactorial nature [1–5]. As the evaluation of drug efficacy lies at the heart of pharmacological research for the benefit of mankind, statistical models that can handle multiple variables are required.

4.2 Theoretical Background of Novel Methodology

Hierarchical cluster analysis is based on the concept that cases (patients) with closely connected characteristics might also be more related in other fields like drug efficacy. With large data it is a computationally intensive method, and today commonly classified as one of the methods of explorative data mining [8]. This is, obviously, explorative rather than confirmative research, but, if you choose your variables carefully, like severely oncogenic genes in patients with cancer, this exercise can be very rewarding, and can be helpful to identify subgroups at particular risks. We should add that other methods for clustering are available, e.g., centroid clusters and density based clusters. The advantage of the first is that it accounts a measure of uncertainty in the clusters, but it is more complex. The advantage of the second is that it is suitable for non-Gaussian data, but on the other hand it is rather structure-less.

4.3 Flexibility of Hierarchical Cluster Analysis

In the current example only 9 variables were included. This number may seem small. But a micro-array of 500 genes is likely to show only a few, 10–15, highly fluorescent spots indicating over-expressed genes, and it does make sense not to take the under-expressed into account. Nonetheless, when using larger gene arrays, hundreds of highly expressed may be observed, but hierarchical cluster analysis can easily handle these numbers of variables. This is an explanatory article of a relatively novel method, and the novel methodology is hard to explain with larger numbers of variables. Hierarchical cluster analysis is currently often classified as a method of machine learning [8], because it is computationally intensive and allows computers to produce with the help of training data predictions models for future data.

4.4 Comparison with Other Machine Learning Methods

Another machine learning method suitable for simultaneous assessment of multiple variables is factor analysis. Factor analysis is a multivariate technique that replaces a large number of variables with a limited number of novel variables that have the largest possible correlation coefficients with the original variables. It is very different from cluster analysis, because variables are clustered instead of cases (patients). Yet, when using factor analysis for analyzing the example used in the current paper, we came to a rather similar result.

We used SPSS.

> Command: Analyze....Dimension Reduction....Factor....enter variables into Variables box....click Extraction....Method: click Principle Components.... mark Correlation Matrix, Unrotated factor solution....Fixed number of factors: enter 3....Maximal Iterations for Convergence: enter 25....Continue....click Rotation....Method: click Varimax....mark Rotated solution....mark Loading Plots....Maximal Iterations: enter 25....Continue....click Scores.... mark Display factor score coefficient matrix ....OK.

Table 15.6 gives the coefficients of the variables that constitute the 2 latent variables, otherwise called the principal components, obtained by slightly rotating both the x- and y-axes of the model. The genes 13–15, 17, 19, 21 have a strong correlation with the latent variable 1 with correlation coefficients larger than 0.7, and the genes 16, 18, 20 with the latent variable 2. When minimizing the SPSS outcome file, and returning to the data, we could observe that SPSS had produced two latent variable-values for each case (patient). When these latent variables were used in a linear regression as predictors of progression free interval, one of them was statistically significant at $p = 0.002$ (Table 15.7). The first latent variable of the factor analysis was also strongly correlated with the cluster membership (Table 15.8). Obviously, the two methods, although methodologically very different, led to somewhat similar results, although at a different levels of significance.

5 Conclusions

1. There is a body of literature indicating that drug efficacy is not a simple monocausal entity, but rather a complex one of a multifactorial nature.
2. Traditional clinical trials can only count a mean result versus control, and is unable to assess multiple factors.

Table 15.6 Factor analysis modeling

Rotated component matrix[a]

	Component	
	1	2
var 13	,919	,307
var 14	,940	,269
var 15	,931	,277
var 16	,176	,900
var 17	,767	,564
var 18	,429	,874
var 19	,731	,669
var 20	,364	,755
var 21	−,729	−,515

The coefficients of the variables that constitute two latent factors, otherwise called components, obtained by slight rotating the both x- and y-axes of the model. The genes 13–15, 17, 19, 21 have a strong correlation with latent factor 1 and the genes 16, 18, 20 with latent factor 2

Extraction method: Principal component analysis

Rotation method: Varimax with Kaiser normalization

[a]Rotation converged in three iterations

Table 15.7 Linear regression with progression free interval as outcome and the latent factors of the factor analysis as predictor

Coefficients[a]

Model		Unstandardized coefficients		Standardized coefficients		
		B	Std. error	Beta	t	Sig.
1	(Constant)	12,818	,494		25,925	,000
	REGR factor score 1 for analysis 1	−2,259	,519	−,800	−4,356	,002
	REGR factor score 2 for analysis 1	−,843	,519	−,299	−1,626	,143

[a]Dependent variable: progressionfreeinterval

3. Hierarchical cluster analysis is able to do so, and performed well in the example given: two clusters of patients were identified with significantly different progression free intervals.

4. Hierarchical cluster analysis may be more appropriate for drug efficacy analyses than other machine learning methods, like factor analysis, because the patients themselves rather than some new characteristics are used as dependent variables.

Table 15.8 Binary logistic regression with cluster membership as outcome and the latent variables as predictors

Variables in the equation							
		B	S.E.	Wald	df	Sig.	Exp(B)
Step 1[a]	FAC1_1	−47,403	17086,437	0,000	1	0,998	0,000
	FAC2_1	−0,788	29179,373	0,000	1	1,000	0,455
	Constant	−1,222	28907,716	0,000	1	1,000	0,295

[a]Variable(s) entered on step 1: FAC1_1, FAC2_1

References

1. Solyanik GI (2010) Multifactorial nature of tumor drug resistance. Exp Oncol 32:181–185
2. Tsao DA, Chang HJ, Hsiung SK, Huang SE, Chang MS, Chiu HH, Chen YF, Cheng TL, Shiu-Ru L (2010) Gene expression profiles for predicting the efficacy of the anticancer drug 5-fluorouracil in breast cancer. DNA Cell Biol 29:285–293
3. Latan MS, Laddha NC, Latani J, Imran MJ, Begum R, Misra A (2012) Suppression of cytokine gene expression and improved therapeutic efficacy of microemulsion -based tacrolimus cream for atopic dermatitis. Drug Deliv Transl Res 2:129–141
4. National Institutes of Health, 9000 Rockville Pike, Bethesda, Maryland. Enhancing HIV vaccine efficacy in high-risk drug users. Release data Jan 6 2003, RFA Number DA-03-002
5. Alsford S, Eckert S, Baker N, Glover L, Sanchez-Flores A, Leung KF, Turner DJ, Field MC, Berriman M, Horn D (2012) High throughput decoding of antitrypanosomal drug efficacy and resistance. Nature. doi:10.1038/nature10771
6. Sibson R (1973) An optimally efficient algorithm for a single-link cluster method. Comput J 16:30–34
7. Defays D (1977) An efficient algorithm for a complete link cluster method. Comput J 20:364–366
8. Anonymous (2012) Cluster analysis. http://en.wikipedia.org/Clustert_analysis. Accessed 12 July 2012
9. Anonymous (2012) Sequence clustering. http://en.wikipedia.org/wiki/Sequence_analysis. Accessed 12 July 2012
10. Kim HK, Choi IJ, Kim HS, Oshima A, Michalowski A, Green JE (2011) A gene expression signature of acquired chemoresistance to cisplatinum and fluorouracil combination chemotherapy in gastric cancer patients. PLoS One 18:e16694
11. Yeh ST (2012) Clinical adverse event data analysis and visualization. Smith Kline Datafile, Accessed 10 July 2012
12. Bychowiec B, Piskorski J, Stanislawska K, Dziarmaga M, Mineczykowski A, Wykretowicz A, Wysocki H (2010) An exploratory clustering study of rare adverse events in drug deluting stent patients. Comput Methods Sci Technol 16:5–11
13. Xu D, Redamn-Furey N (2007) Statistical cluster analysis of pharmaceutical solvents. Int J Pharm 339:175–188
14. Hawwa A, Millership JS, Collier PS, McCarthy A, Dempsey S, Cairns C, McElnay JC (2009) Development of objective methodology to measure medication adherence to oral thiopurines in paediatric patients with acute lymphoblastic leucemia. Eur J Clin Pharm 65:1105–1112
15. SPSS statistical software (2012) www.spss.com. Accessed 12 July 2012

Chapter 16
Partial Least Squares

1 Summary

1.1 Background

Traditional statistical tests are unable to handle a large number of variables. The simplest method to reduce large numbers of variables is the use of add-up scores. But add-up scores do not account for the relative importance of the separate variables, their interactions and differences in units. Principal components analysis and partial least square analysis account all of that, but are virtually unused in clinical trials.

1.2 Objective

To assess the performance of either of the two methods.

1.3 Methods

A simulated example of 250 patients' gene expression data as predictor and drug efficacy scores as outcome will be used. For principal components analysis SPSS's Data Dimension Reduction module was used, for partial least square analysis R Partial Least Squares, a free statistics and forecasting software was used.

T.J. Cleophas and A.H. Zwinderman, *Machine Learning in Medicine*,
DOI 10.1007/978-94-007-5824-7_16, © Springer Science+Business Media Dordrecht 2013

1.4   Results

Of 27 variables three novel predictor variables were constructed. With principal components analysis the three were very significant predictors of the add-up outcome score with t-values of 10.2, 21.6, and 6.7 ($p < 0.000$, $p < 0.000$, $p < 0.000$). Partial least squares included the outcome variables in its program, and was also able to predict the outcome variables although at a lower level of significance with t-values of 6.8, 16.2, and 3.5 ($p < 0.000$, $p < 0.000$, $p < 0.001$). Traditional multiple linear regression with the novel predictors in the form of add-up scores as independent variables produced a consistent further reduction of significance with t-values of 3.4, 11.2 and 2.4 ($p < 0.002$, $p < 0.001$, $p < 0.02$).

1.5   Conclusions

1. Principal components analysis and partial least squares can handle many more variables than the standard covariance methods like MANOVA and MANCOVA can, and is more sensitive than add-up scores.
2. The methods account the relative importance of the separate variables, their interactions and differences in units.
3. They are also very flexible, to the extent that manifest variables can be applied twice, first in the form of clusters for prediction and second unclusteredly as manifest outcome variables.
4. Partial least squares method is parsimonious to principal components analysis, because it can include outcome variables in the model.

2   Introduction

Current clinical trials tend to include large numbers of variables. For example, a series of gene expressions can be used to predict the efficacy of cytostatic treatment [1, 2], repeated measurements can be used as endpoint in randomized longitudinal trials [3, 4], and multi-item personal scores can be used for the evaluation of antidepressants [5]. Many more examples can be given. Some kind of mathematical modeling of the multiple variables is required for useful information. Often multiple linear models like MANOVA (multivariate analysis of variance) and MANCOVA (multivariate analysis of covariance) are successful and can handle thousands of cases. However, the models have great difficulty to model more than two or three dependent variables. The simplest method to reduce large numbers of variables is the use of add-up scores. But add-up scores do not account for the relative importance of the separate variables, their interactions and differences in units. All of this is accounted for by a technique, developed in the early 1950s by the psychometrician Eastman (London UK), called principal components analysis: two or three unmeasured factors, otherwise called components or latent variables (LVs), are identified to

Table 16.1 Data dimension reduction with principal components and partial least square models

Principal components analysis		Partial least squares analysis			
Predictor MVs	Predictor LVs	Predictor MVs	Predictor LVs	Outcome LVs	Outcome MVs
1 →	1	1 →	1		
2 →		2 →			
3 →		3 →			
4 →		4 →			
				→ 3	← 9
5 →	2	5 →	2		← 10
6 →		6 →			← 11
7 →		7 →			
8 →		8 →			

Both are based on multivariate linear regression. The arrows are the fit linear correlation coefficients as calculated by the software. Compared to the principal components models, the partial least square models are mathematically more complex, because they calculate, in addition to the best fit predictor LVs, the best fit outcome LVs, and make for that purpose use of both predictor and outcome MVs. The latter models, because they take more into account, will produce smaller test statistics, but are at the same time less at risk of biases, e.g., due to differences in importance of manifest outcome variables, their interactions and differences in units

explain a much larger number of measured variables, otherwise called manifest variables (MVs) [6, 7]. Partial least squares analysis is an extension of principal components analysis, first described by the econometrist Wold, Stockholm 1966, and is appropriate for estimating more than a single layer of latent variables [8].

Although both methods have become major research tools in behavioral sciences, social sciences, marketing, operational research, and other applied sciences [9–11], they are rarely applied in clinical trials. When searching the internet we found, except for a few genetic trials [12–15], virtually no clinical studies. This is a pity given the presence of large numbers of variables in this field of science.

In the current chapter we will assess the performance of either of the two methods. A simulated example of 250 patients' gene expression data and drug efficacy scores will be used.

3 Some Theory

Examples of principal components and partial least square models are in Table 16.1. Both are based on multivariate linear regression. The arrows are the fitted linear correlation coefficients. Compared to the principal components models, the partial least square models are mathematically more complex, because they calculate, in addition to the best fit predictor LVs, the best fit outcome LVs, and make for that purpose use of both predictor and outcome MVs. The latter models, because they take more into account, will produce smaller test statistics, but are, at the same time, less at risk of biases, e.g., due to differences in importance of manifest outcome variables, their interactions and differences in units.

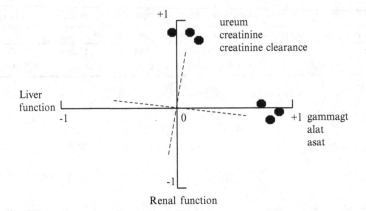

Fig. 16.1 The relationships of six manifest variables with two latent variables, the liver and renal function as linearly modeled by principal components analysis. It can be observed that ureum, creatinine and creatinine clearance have a very strong positive correlation with renal function and the other three with liver function. It can be demonstrated that by slightly rotating both x and y-axes the model can be fitted even better

3.1   Principal Components Analysis

Principal components analysis (Table 16.1) uses all manifest variables available to linearly model the best fit correlation coefficients between the MVs and the LVs along an x and y-axis. The magnitude of a LV is calculated as the weighted mean of all of the MVs along one axis. Figure 16.1 gives a simple example with renal and liver function as latent variables. It can be observed that ureum, creatinine and creatinine clearance have a very strong positive correlation with renal function and the other three with liver function. It can be demonstrated that by slightly rotating both x and y-axis the model can be fitted even better. If a third latent variable existed within a data file, it could be represented by a third axis, a z-axis creating a 3-d graph. Also additional factors can be added to the model, but they cannot be presented in a 2- or 3-d drawing, but, just like with multiple regression modeling, the software programs have no problem with multidimensional calculations similar to the above 2-d calculations.

3.2   Partial Least Squares Analysis

Partial least squares analysis (Table 16.1) does not simultaneously use all of the predictor variables available, but rather uses an a priori clustered set of predictor variables of 4 or 5 to calculate a new latent variable. Unlike principal components analysis which does not consider response variables at all, partial least square analysis does take response variables into account and therefore often leads to a better fit of the response variable. Correlation coefficients are produced from multivariate linear regression rather than fitted correlation coefficients along the x and y-axes.

4 Example

A 250 patients' data-file (Appendix) was supposed to include 27 variables consistent of both patients' microarray gene expression levels and their drug efficacy scores. All variables were standardized by scoring them on an 11 points linear scale (0–10). The following clusters of genes were highly correlated with one another: the variables 1–5, the variables 16–19, and the variables 24–27. The variables 20–23 were supposed to represent drug efficacy scores and were clustered as the outcome variables. The data file is given in the Appendix.

4.1 Principal Components Analysis

For principal components analysis SPSS's Data Dimension Reduction module [7] was used. First the reliability of the model was assessed by testing the test-retest reliability of the original variables. The test-retest reliability of the original variables should be assessed with Cronbach's alphas using the correlation coefficients after deletion of one variable: all of the data files should produce at least by 80% the same result as that of the non-deleted data file (alphas >80%).
Command: Analyze....Scale....Reliability Analysis....transfer original variables to Variables box....click Statistics....mark Scale if item deleted....mark Correlations.... Continue....OK.

The Table 16.2 shows, that, indeed, none of the original variables after deletion reduces the test-retest reliability. The data are reliable. We will now perform the principal components analysis with three components, otherwise called latent variables.
Command: Analyze....Dimension Reduction....Factor....enter variables into Variables box....click Extraction....Method: click Principle Components....mark Correlation Matrix, Unrotated factor solution....Fixed number of factors: enter 3....Maximal Iterations plot Convergence: enter 25....Continue....click Rotation.... Method: click Varimax....mark Rotated solution....mark Loading Plots....Maximal Iterations: enter 25....Continue....click Scores.... mark Display factor score coefficient matrix ....OK.

The Table 16.3 shows the best fit coefficients of the original variables constituting three components. The component 1 has a very strong correlation with the variables 16–19, the component 2 with the variables 24–27, and the component 3 with the variables 1–4.

These three components can, thus, be interpreted as the latent predictor variables. When minimizing the outcome file and returning to the data file, we now observe, that, for each patient, the software program has produced the individual values of these novel predictors.

Table 16.2 Validation of the data: there should be a strong correlation between the scores within the clusters (strong collinearity)

Item-total statistics

	Scale mean if item deleted	Scale variance if item deleted	Corrected item-total correlation	Squared multiple correlation	Cronbach's alpha if item deleted
VAR00001	79,6200	277,273	,547	,486	,903
VAR00002	79,6200	263,980	,724	,700	,896
VAR00003	79,5120	264,749	,743	,671	,895
VAR00004	79,5480	284,361	,477	,385	,906
VAR00005	81,0720	264,501	,566	,386	,903
VAR00016	80,3720	257,166	,714	,623	,895
VAR00017	79,7320	268,494	,665	,582	,898
VAR00018	80,3080	265,869	,588	,477	,902
VAR00019	80,9560	255,038	,719	,555	,895
VAR00024	80,2800	245,696	,719	,611	,895
VAR00025	80,0200	272,702	,507	,340	,905
VAR00026	80,6240	244,581	,714	,627	,896

The test-retest reliability of the variables is assessed with Cronbach's alphas. All of the reliability assessments should produce at least for 80% the same result (Cronbach's alphas >80%)

Table 16.3 Principal components analysis

Rotated component matrix[a]

	Component		
	1	2	3
VAR00001	,249	,136	,797
VAR00002	,582	,128	,652
VAR00003	,616	,163	,586
VAR00004	,003	,364	,770
VAR00005	,711	,063	,211
VAR00016	,819	,242	,127
VAR00017	,500	,516	,217
VAR00018	,379	,482	,306
VAR00019	,719	,289	,235
VAR00024	,585	,617	,079
VAR00025	,160	,675	,228
VAR00026	,634	,563	,050
VAR00027	,084	,823	,172

Component 3 has a strong positive correlation with the MVs 1–4, component 2 with MVs 24–27, and component 1 with MVs 16–19
Extraction method: principal component analysis
Rotation method: varimax with Kaiser normalization
[a]Rotation converged in eight iterations

Table 16.4 A multiple linear regression was performed with the add-up scores of the outcome variables instead of MANOVA

Coefficients[a]

Model		Unstandardized coefficients		Standardized coefficients	t	Sig.	Collinearity statistics	
		B	Std. Error	Beta			Tolerance	VIF
1	(Constant)	27,364	,231		118,420	,000		
	REGR factor score 1 for analysis 1	4,991	,232	,735	21,558	,000	1,000	1,000
	REGR factor score 2 for analysis 1	2,358	,232	,347	10,185	,000	1,000	1,000
	REGR factor score 3 for analysis 1	1,552	,232	,229	6,701	,000	1,000	1,000

SPSS statistical software did not execute the command for MANOVA, and explained that too many columns and too many levels were in the data for the purpose of a proper analysis
[a]Dependent variable: 20–23

In order to fit these novel predictors with the outcome variables, the drug efficacy scores (variables 20–23), multivariate analysis of variance (MANOVA) should be appropriate given the continuous nature of the four outcome variables. However, the large number of columns in the design matrix caused integer overflow, and the large number of columns caused too many levels within some components as well as numerical problems with higher order interactions among components, and the command was not executed. Instead we performed univariate multiple linear regression with the add-up scores of the outcome variables as novel outcome variable. Table 16.4 gives the results. All of the three latent predictors were very significant independent predictors of the add-up outcome variable.

4.2 Partial Least Squares Analysis

Because partial least is not available in the basic and regression modules of SPSS, we used the software program R Partial Least Squares, a free statistics and forecasting software available on the internet as a free online software calculator [16]. The data-file was imported directly from a Word file. The selected clusters of variables were listed: latent variable 2 (16–19), latent variable 1 (24–27), latent variable 4 (1–4), and latent outcome variable 3 (20–23).

A square boolean matrix was constructed with "0 or 1" if fitted correlation coefficients were to be included in the model "no or yes". Then the order "compute" was given.

Latent variable				
Latent variable	1	2	3	4
1	0	0	0	0
2	0	0	0	0
3	1	1	0	0
4	0	0	1	0

After 15 s of computing the program produced the following results. First, the data were validated using the GoF (goodness of fit) criteria. GoF=√ [mean of r-square values of comparisons in model * r-square overall model], where * is the sign of multiplication. A GoF value varies from 0 to 1 and values larger than 0.8 indicate that the data are adequately reliable for modeling. The following results were given:

GoF value	
Overall	0.9459
Outer model (including manifest variables)	0.9986
Inner model (including latent variables)	0.9466

The data were, thus, adequately reliable. The calculated best fit r-values (correlation coefficients) were estimated directly from the model, and their standard errors would be available from second derivatives. However, the problem with the second derivative procedure is that it requires very large data files in order to be accurate. Instead of an inaccurate estimate of the standard errors, distribution free standard errors were calculated using bootstrap resampling.

Latent variables	Original r-values	Bootstrap r-values	Standard error
1 versus 3	0.57654	0.57729	0.08466
2 versus 3	0.67322	0.67490	0.04152
4 versus 3	0.18322	0.18896	0.05373

The Table 16.5 shows that all of the three correlation coefficients were very significant predictors, and that the three predictor latent variables were, thus, very significant predictors of the latent outcome variable.

4.3 Comparison of the Two Models

Table 16.5 shows the correlation coefficients of predictor latent variables in either of the two models used in this paper. The partial least square method produced somewhat smaller test statistics, but it is less biased because it takes into account the relative importance of the separate manifest variables, and their interactions. In spite of accounting more, the level of statistical significance of the latter model remained excellent.

Table 16.5 Comparison of correlation coefficients of predictor latent variables

Principal components		Partial least squares		Add-up scores	
Correlation coefficients	t-value	Correlation coefficients	t-value	Regression coefficients	t-value
0.74	10.2	0.58	6.8	0.15	3.4
0.35	21.6	0.67	16.2	0.61	11.2
0.23	6.7	0.19	3.5	0.14	2.4

The partial least square method produced somewhat smaller t-values, but it is less biased and the level of statistical significance is excellent even so. When using the add-up scores of the main variables of the three components instead of the modeled latent variables, the effects were similarly statistically significant, but, the magnitudes of the t-values further fell

When using the add-up scores of the main variables of the three components instead of the modeled latent variables, the effects were similarly statistically significant. However, they were so at lower levels of significance (Table 16.5). Obviously, the principal components and partial least squares analyses provided better fit for the data than did multiple linear regression with add-up variables as both predictor and outcome variables.

5 Discussion

The data dimension reduction methods explained in this paper are wonderful for the analysis of data with many variables, because they can handle many more variables than the standard covariance methods like MANOVA and MANCOVA can. They also have the advantage, compared to models using the composite of multiple variables as endpoint or predictor, that they can account for the relative importance of the separate variables, their interactions and differences in units.

However, the data dimension reduction methods do have a numbers of limitations. They do not comply with all of the requirements of normal distributions in the data, adequate sample sizes to reduce type II errors, adjustments for multiple testing to reduce type I errors etc. Also, they are scientifically less rigorous than traditional methods, because empirical rather than parametric confidence intervals are applied and hypothesis testing is based on re-sampling methods such as jack-knife and bootstrap. However, a reduced scientific rigor is equally true for most traditional multiple variables and multivariate analyses, particularly if they are post hoc and not based on prior hypotheses. We should add that, particularly, with sound underlying clinical arguments, the novel methodologies are helpful for confirm hypotheses, increasing precision of some point estimates, benefit risk analyses, providing relevant arguments for clinical decision making, and other quantitative assessments.

A special advantage of the novel methodologies is their flexibilities, and capacities to handle numerous variables. Forty variables or even more is no problem [17]. If you don't have outcome variables, the two layers of a two-layer partial least squares model can be constructed using the manifest variables twice, first in the form of clusters for prediction and second unclusteredly as manifest outcome variables [17].

The principal components analysis does not consider response variables, and partial least square analysis does take response variables into account and therefore often leads to a better fit of the response variable. Correlation coefficients are produced from multivariate linear regression rather than fitted correlation coefficients along the x and y-axes. The latter method may be parsimonious to the former, that is, if outcome variables are in your data.

At this time the novel models can not yet be applied for binary data like survival data, but this is a matter of time. Bastien (Aulnay, France) has already proposed a PLS-Cox model for the analysis of the effect of gene expression on survival [14]. Also non linear models do not fit the novel methods. Outlier identification with linear models makes use of tests based on normal distribution and homoscedasticity assumptions like the Durbin Watson test and is not available with the novel methods. And, so, despite the pleasant properties of the novel methods, there is plenty room for improvement.

6 Conclusions

Advantages of the novel methods include

1. they can handle many more variables than the standard covariance methods like MANOVA and MANCOVA can, and are more sensitive than add-up scores are,
2. they account the relative importance of the separate variables, their interactions and differences in units,
3. they are very flexible, to the extent that manifest variables can be applied twice, first in the form of clusters for prediction and second unclusteredly as manifest outcome variables.

Limitations of the novel methods include

1. they do not comply with the normal distribution and homoscedasticity,
2. they are at increased risk of type II errors,
3. they are at increased risk of type I errors.

Partial least squares method is parsimonious to principal components analysis because it can include outcome variables in the model.

There is room for improvement of the novel methods, because, to date,

1. binary variables cannot be included,
2. non linear variables cannot be included,
3. no tests for outliers is included.

7   Appendix: Datafile of the Example Used in the Present Chapter

G1	G2	G3	G4	G16	G17	G18	G19	G24	G25	G26	G27	O1	O2	O3	O4
8	8	9	5	7	10	5	6	9	9	6	6	6	7	6	7
9	9	10	9	8	8	7	8	8	9	8	8	8	7	8	7
9	8	8	8	8	9	7	8	9	8	9	9	9	8	8	8
8	9	8	9	6	7	6	4	6	6	5	5	7	7	7	6
10	10	8	10	9	10	10	8	8	9	9	9	8	8	8	7
7	8	8	8	8	7	6	5	7	8	8	7	7	6	6	7
5	5	5	5	5	6	4	5	5	6	6	5	6	5	6	4
9	9	9	9	8	8	8	8	9	8	3	8	8	8	8	8
9	8	9	8	9	8	7	7	7	7	5	8	8	7	6	6
10	10	10	10	10	10	10	10	10	8	8	10	10	10	9	10
2	2	8	5	7	8	8	8	9	3	9	8	7	7	7	6
7	8	8	7	8	6	6	7	8	8	8	7	8	7	8	8
8	9	9	8	10	8	8	7	8	8	9	9	7	7	8	8
7	7	8	8	8	9	10	7	9	4	8	8	9	8	7	7
3	4	3	8	4	4	4	3	4	3	4	4	4	4	3	4
7	8	8	5	8	8	7	6	7	7	8	7	10	8	8	7
8	8	8	8	6	8	5	1	9	7	7	8	7	7	8	6
7	8	8	8	8	9	8	7	10	10	9	8	9	9	9	9
8	4	3	8	3	5	5	3	2	10	1	0	5	3	4	3
8	7	6	10	8	8	7	6	4	4	5	5	7	7	7	5
9	9	10	8	8	9	7	7	8	9	8	9	8	7	8	7
6	6	6	6	4	5	4	5	3	9	3	4	4	5	4	3
8	8	8	7	7	7	8	6	8	7	9	4	6	7	8	9
9	9	10	9	10	10	7	10	10	10	10	10	8	8	8	5
8	7	8	8	9	8	9	8	8	8	8	8	8	8	8	9
8	5	5	4	2	1	1	0	0	1	0	0	3	2	4	5
6	6	6	6	5	6	3	5	4	4	4	5	5	6	3	4
7	8	9	8	8	9	9	6	9	8	8	10	9	8	7	7
8	8	8	7	7	7	7	6	7	8	7	8	7	6	6	6
8	8	8	8	9	8	9	8	9	8	9	9	9	8	7	8
7	7	7	6	7	7	9	7	7	7	7	8	8	6	7	7
9	9	9	9	6	9	8	7	8	8	8	9	8	8	8	8
10	10	10	10	9	9	10	5	10	2	9	9	8	10	8	8
9	8	9	9	8	7	7	8	9	9	9	9	8	5	9	7
8	9	9	9	8	7	7	6	7	8	8	8	8	7	8	6
3	4	2	5	4	2	2	4	4	4	3	4	6	2	3	2
8	8	9	9	8	8	8	8	8	8	8	8	8	8	8	8
8	6	7	6	7	7	8	6	7	6	5	5	6	7	7	6
10	10	10	10	7	10	10	8	10	10	10	10	10	8	8	7
8	10	9	8	8	8	7	6	7	7	10	8	9	8	8	7
8	8	8	8	8	8	7	8	7	8	8	8	9	9	8	7
5	7	7	8	5	7	7	3	1	6	3	10	5	6	6	5
10	9	9	10	7	9	9	9	9	9	9	8	8	9	7	7
9	7	7	9	3	6	4	2	1	8	2	1	6	6	6	6

(continued)

(continued)

G1	G2	G3	G4	G16	G17	G18	G19	G24	G25	G26	G27	O1	O2	O3	O4
8	8	10	8	9	8	7	8	8	7	8	8	9	6	5	7
6	8	8	8	9	10	10	9	10	9	9	10	9	8	5	5
8	8	8	8	10	8	7	10	8	8	7	10	9	7	8	6
6	5	5	6	6	6	4	6	3	5	0	3	7	5	5	3
9	9	9	8	8	9	8	7	6	7	8	10	8	8	8	6
9	10	8	8	9	10	10	9	7	8	9	7	8	8	7	7
8	8	8	9	6	8	7	6	8	9	8	8	7	7	6	5
8	5	6	7	8	8	7	7	4	6	7	6	8	8	7	6
4	1	4	9	0	0	7	0	0	10	0	10	0	0	0	0
5	5	7	5	7	7	8	5	7	7	5	5	7	7	7	7
5	5	6	5	4	4	4	3	3	2	3	3	3	4	3	3
7	9	9	10	5	9	9	9	9	6	7	6	10	7	10	9
10	10	10	10	8	9	9	6	7	8	8	10	7	7	7	6
8	8	8	8	6	9	8	7	7	6	6	2	7	7	7	5
6	6	7	9	8	8	7	6	1	9	0	4	6	7	7	6
6	7	7	7	6	5	5	5	5	7	3	5	7	6	6	8
9	9	9	9	8	8	9	6	8	7	6	10	8	7	7	8
7	7	7	7	6	8	8	6	7	7	7	8	6	6	5	10
9	7	8	9	8	10	8	9	8	9	7	9	7	7	8	3
8	9	9	8	7	8	7	8	8	6	7	8	7	8	7	6
8	8	8	8	6	8	8	5	8	9	8	7	7	7	6	5
7	7	7	7	4	5	6	6	3	6	7	7	1	5	6	5
9	10	9	9	8	9	8	8	9	8	9	9	8	7	8	8
8	9	9	8	8	8	8	7	8	7	8	9	6	6	5	6
7	8	8	8	6	7	7	6	8	5	7	7	7	6	7	5
4	2	2	6	5	5	4	4	6	4	3	2	4	6	7	2
5	5	7	5	5	5	5	2	2	9	5	5	4	5	5	4
9	9	10	9	7	8	7	8	8	9	8	8	8	8	6	9
8	8	8	8	7	7	7	9	8	9	7	8	7	7	5	6
8	8	9	8	8	9	5	9	8	5	7	6	8	8	8	6
9	9	9	9	6	8	8	4	7	5	6	6	7	7	8	8
9	8	8	8	7	9	9	9	10	10	10	10	10	9	7	10
9	9	9	8	8	8	8	7	7	7	7	7	8	8	8	7
8	5	7	9	2	8	8	2	9	10	1	9	5	5	5	5
7	6	9	8	5	7	7	6	5	7	4	4	6	7	6	7
8	8	9	8	6	7	7	6	8	7	7	10	8	7	8	6
10	10	10	10	8	10	10	7	8	8	7	8	9	9	9	7
9	9	6	6	4	5	5	5	2	3	5	4	2	3	3	3
3	3	3	8	0	7	0	0	0	7	0	10	0	0	8	8
5	4	4	7	4	4	4	2	0	4	2	8	3	3	3	3
8	10	10	10	7	8	7	10	10	9	8	10	10	9	9	8
5	8	8	8	7	8	8	6	7	7	7	10	7	8	6	6
7	4	5	9	5	8	7	5	5	8	0	7	6	6	6	6
5	6	5	8	10	9	0	8	8	8	8	5	8	8	5	4
7	5	7	6	3	6	6	3	5	6	6	5	5	5	5	5
10	8	9	8	8	8	8	6	8	8	6	6	8	7	5	8
10	10	10	10	10	10	8	10	9	10	10	10	10	9	9	8

(continued)

(continued)

G1	G2	G3	G4	G16	G17	G18	G19	G24	G25	G26	G27	O1	O2	O3	O4
6	6	4	5	0	5	5	5	5	8	5	9	6	4	5	5
10	3	7	9	0	5	7	7	10	8	10	10	5	5	5	5
5	7	8	7	8	7	8	7	8	6	7	6	8	6	7	6
9	10	9	9	10	6	6	7	9	8	8	8	10	7	7	10
10	10	10	10	9	10	10	9	10	10	10	10	9	9	9	9
10	10	10	10	10	10	10	8	10	10	6	10	7	6	8	8
7	7	7	8	7	8	8	6	8	8	7	7	8	7	8	8
9	5	7	9	6	8	8	4	6	7	4	5	6	5	5	4
9	9	10	8	8	9	8	7	8	8	7	9	8	7	5	7
6	6	5	4	4	4	4	3	4	3	4	5	4	5	5	5
7	8	8	9	7	5	4	7	10	8	8	8	6	4	4	7
8	6	6	5	7	6	0	8	7	9	7	7	7	7	6	7
6	8	8	9	9	9	9	5	9	8	7	9	9	5	5	9
9	5	6	7	10	10	8	7	8	9	10	10	8	8	7	8
8	7	8	5	8	7	4	5	8	5	5	9	3	5	3	5
7	8	7	4	8	8	8	7	7	6	6	7	8	7	7	7
8	7	10	10	10	10	10	10	10	10	10	10	9	6	4	8
5	9	10	5	9	9	6	8	10	10	10	9	8	7	9	9
9	6	6	7	10	10	6	6	9	10	10	9	10	10	10	9
0	4	7	5	10	8	9	9	9	7	8	7	8	9	9	9
4	8	8	6	9	9	7	2	9	9	9	9	8	8	8	7
8	8	10	8	7	7	5	5	5	10	8	3	7	7	6	7
9	10	10	7	5	4	0	7	10	10	10	10	5	4	5	9
10	10	10	10	7	0	0	8	2	8	1	0	4	5	3	3
10	8	8	8	5	5	8	8	10	10	10	10	6	6	5	5
7	10	10	8	10	10	8	8	10	9	10	10	7	8	10	6
10	9	9	6	9	9	0	9	10	8	9	9	8	7	10	7
8	10	8	5	7	6	5	7	10	10	10	10	6	6	7	7
10	8	8	7	8	8	7	5	10	8	8	10	8	8	7	8
8	7	8	8	10	10	2	1	8	10	8	8	9	7	9	10
8	8	8	8	6	7	7	4	8	8	7	7	7	5	6	7
7	9	8	8	9	8	8	7	9	9	9	7	10	9	7	7
8	8	9	9	7	7	8	7	7	8	7	7	7	7	8	8
8	7	8	7	8	8	8	7	8	8	7	8	8	8	8	7
8	7	7	8	7	7	8	7	8	8	7	8	7	7	8	7
8	8	8	8	7	6	8	6	9	8	7	9	8	8	6	6
8	8	8	9	9	6	8	9	8	9	10	10	8	8	8	5
7	8	8	6	8	9	9	6	8	8	8	8	8	8	6	8
7	9	9	8	6	8	8	5	8	7	5	9	7	5	7	4
10	10	10	8	9	8	8	8	10	10	10	10	10	10	9	9
6	8	7	8	9	8	10	8	8	9	9	8	8	7	7	5
8	8	8	8	8	8	8	8	8	8	5	10	8	8	8	7
10	0	0	10	0	7	5	0	0	3	0	10	0	0	0	0
8	5	9	4	6	8	8	5	6	6	4	5	6	5	5	4
9	9	9	9	8	8	8	7	7	3	0	9	7	7	8	8
8	9	8	8	8	8	8	8	8	9	9	8	8	8	9	5
7	7	7	7	7	7	7	5	7	7	7	5	8	7	5	6

(continued)

(continued)

G1	G2	G3	G4	G16	G17	G18	G19	G24	G25	G26	G27	O1	O2	O3	O4
9	9	9	9	7	7	8	8	8	7	8	6	8	6	6	7
5	7	4	10	0	10	10	0	5	5	0	10	0	0	0	0
9	9	9	9	9	10	10	9	10	10	10	10	10	10	5	5
8	8	9	7	7	8	8	7	8	7	7	8	8	8	6	8
9	10	10	7	9	9	8	4	9	9	9	8	8	7	9	9
10	10	10	10	10	10	9	7	10	10	10	9	7	7	5	9
8	6	9	9	7	9	8	5	6	6	5	5	6	7	5	4
7	7	8	5	8	8	7	6	5	5	7	4	5	6	6	6
9	10	10	10	9	8	9	8	8	8	8	9	9	8	6	7
7	7	6	6	4	6	6	4	4	6	3	5	4	4	4	4
8	8	8	8	9	8	7	9	10	3	7	10	9	8	7	7
8	8	8	8	7	8	5	8	10	10	7	10	8	7	7	7
10	10	10	10	10	10	10	10	10	10	10	9	10	10	10	10
10	10	10	10	9	10	10	9	10	10	10	10	9	9	9	8
9	10	10	10	8	10	10	8	10	10	10	10	9	8	8	7
4	6	8	8	7	7	7	5	4	7	5	9	6	6	7	5
8	8	8	7	7	8	9	7	7	5	7	4	8	9	9	9
8	8	8	8	6	7	7	4	6	10	6	6	7	7	7	5
8	8	4	8	5	5	5	1	0	5	0	10	2	2	2	2
7	7	7	7	7	8	8	4	7	7	6	6	6	6	6	6
8	7	7	8	10	9	8	9	10	9	8	9	9	8	7	8
9	9	7	8	9	8	8	8	8	8	9	8	9	7	8	6
5	3	4	3	4	5	3	5	2	3	5	4	4	2	4	7
6	8	8	8	9	9	8	7	9	8	9	10	8	8	7	7
9	10	10	10	6	8	9	8	0	10	10	10	10	9	6	9
4	5	5	7	4	4	5	4	2	4	2	7	5	5	3	3
8	8	8	8	10	10	10	10	10	10	10	8	10	7	7	7
9	9	9	9	10	8	8	8	8	8	7	8	9	9	8	8
10	10	10	10	8	8	8	8	8	8	8	9	9	9	8	8
10	10	10	9	10	10	10	10	10	10	10	10	10	10	10	10
10	10	10	10	7	5	5	5	6	8	8	5	8	5	5	10
7	8	8	8	4	5	5	4	5	4	5	8	7	6	8	4
8	8	8	8	5	8	8	5	5	5	5	7	6	6	5	5
8	6	8	5	5	5	5	3	3	9	3	2	5	3	5	3
10	10	10	10	10	10	10	10	9	10	10	10	10	9	10	10
7	7	7	7	7	8	8	5	6	7	7	9	6	7	5	5
8	7	7	8	8	9	5	5	6	7	6	5	7	7	6	6
10	10	10	10	9	10	10	10	9	10	10	10	10	9	10	5
7	9	9	9	8	9	8	8	9	8	8	7	9	10	8	8
9	8	8	8	9	9	8	7	10	8	9	10	9	8	7	8
8	6	6	7	5	7	5	4	5	2	5	5	6	5	5	4
8	9	9	9	6	8	7	6	6	5	5	7	7	6	7	6
7	8	9	9	9	10	10	7	10	5	8	8	10	10	5	9
9	8	8	8	8	9	7	8	0	5	7	10	8	8	9	2
10	10	10	10	6	10	7	8	10	9	2	8	9	9	7	6
10	10	9	10	10	10	10	9	10	10	10	10	10	10	9	10
8	9	9	8	8	8	8	8	8	8	8	8	8	8	9	8

(continued)

(continued)

G1	G2	G3	G4	G16	G17	G18	G19	G24	G25	G26	G27	O1	O2	O3	O4
8	10	10	10	8	8	8	8	9	9	9	8	9	8	9	8
8	8	8	5	5	8	8	8	6	8	10	5	7	7	5	7
6	6	7	7	6	7	5	2	5	5	5	0	6	10	6	6
10	10	10	10	5	10	10	10	10	10	10	10	10	10	5	10
8	7	8	8	7	9	9	7	6	8	8	8	7	7	5	6
8	7	8	7	8	8	8	8	9	9	8	9	8	7	7	6
7	7	7	8	8	9	8	7	8	8	8	9	7	7	7	7
10	10	10	10	10	10	10	10	10	10	10	10	10	10	10	10
10	10	10	9	7	9	9	7	8	8	8	7	8	8	8	8
10	10	10	10	10	10	10	5	10	10	10	10	9	10	9	9
10	10	10	10	10	10	10	10	10	10	10	10	10	10	9	9
10	10	10	9	10	10	9	9	10	6	10	10	10	10	7	9
7	9	9	8	9	10	9	8	8	8	8	8	8	7	5	7
9	9	9	9	9	9	8	8	9	9	8	7	9	8	8	8
6	5	5	7	1	5	6	5	5	10	5	10	3	0	5	5
10	10	10	10	7	10	10	10	10	10	10	10	10	10	5	10
8	9	10	9	9	10	9	9	9	10	10	9	10	9	10	9
6	8	8	9	3	8	5	5	5	5	7	6	5	5	6	6
9	9	9	9	5	8	5	6	9	9	8	10	8	8	8	8
8	9	9	8	5	8	8	8	8	8	7	9	7	7	5	7
6	7	7	7	6	6	6	3	3	6	0	6	5	5	5	5
8	8	8	9	7	8	8	8	5	8	7	10	7	7	7	6
8	8	9	6	6	7	5	5	10	5	0	10	7	7	5	5
8	9	9	7	6	7	7	6	9	7	7	7	7	6	7	7
8	4	6	7	3	6	6	6	0	6	0	9	6	5	4	6
9	9	9	9	9	8	8	8	7	8	8	8	8	8	8	8
6	7	7	6	6	6	4	4	5	6	8	5	2	3	3	4
6	7	7	7	4	6	4	4	4	8	4	5	6	7	7	5
8	7	7	9	7	10	5	6	8	8	6	9	6	7	6	7
10	10	10	9	8	7	8	7	8	9	9	8	5	5	5	4
8	7	8	10	8	9	6	7	8	7	8	8	8	8	7	8
8	9	7	8	9	8	8	7	8	7	5	9	6	8	8	8
7	7	5	7	8	8	6	6	9	7	8	8	7	7	6	7
9	9	10	8	8	8	6	5	10	10	10	10	7	7	5	6
8	6	9	9	8	9	8	9	8	7	7	8	9	9	7	8
7	7	8	9	7	7	7	8	7	8	9	7	6	8	7	7
7	7	8	7	8	7	8	7	8	8	6	5	7	8	7	7
9	10	9	9	8	7	9	9	6	6	6	6	7	9	8	8
7	7	7	6	6	6	9	9	8	3	5	8	6	9	9	8
9	10	7	8	7	5	10	10	10	10	7	10	6	8	9	7
4	6	5	7	4	4	4	3	10	9	10	9	4	4	3	4
8	8	8	8	8	8	7	6	10	8	10	10	10	8	8	7
8	6	3	8	6	8	5	1	10	7	10	10	7	7	8	6
7	7	8	7	8	8	8	6	4	6	5	9	7	7	5	6
9	10	9	9	8	7	7	9	7	4	7	4	5	9	8	6
10	10	8	9	7	6	5	8	6	7	6	6	7	9	9	5
4	6	9	8	9	9	7	7	0	1	0	0	8	7	9	4

(continued)

(continued)

G1	G2	G3	G4	G16	G17	G18	G19	G24	G25	G26	G27	O1	O2	O3	O4
8	8	7	7	4	6	5	5	4	4	4	5	3	5	4	3
8	8	8	8	7	7	8	6	8	7	8	9	6	7	7	9
8	8	3	4	9	8	7	6	6	8	6	8	7	8	8	5
6	8	8	7	5	7	7	7	8	7	8	8	7	7	7	7
9	10	9	8	7	8	8	8	6	7	6	8	8	8	8	8
4	5	5	7	6	9	6	8	9	9	8	8	8	10	10	7
6	8	8	8	8	8	8	6	5	4	6	6	9	9	9	8
9	9	7	7	4	6	5	3	3	5	3	6	4	3	7	5
8	9	5	5	7	8	7	6	6	7	6	6	7	6	7	4
10	10	8	7	7	8	6	5	6	8	5	7	5	6	8	8
7	9	8	7	7	8	10	8	9	8	7	8	9	8	8	7
9	9	9	9	6	3	4	4	7	8	6	8	6	3	4	4
6	5	5	7	6	7	6	4	9	8	8	9	7	7	7	6
9	9	9	9	9	10	10	8	7	6	8	8	8	8	8	7
8	9	9	7	9	9	8	8	3	5	3	6	7	6	7	8
8	8	8	8	4	5	5	8	8	2	8	7	8	7	8	6
8	9	9	4	8	9	9	7	6	7	8	6	9	7	8	7
10	9	7	7	7	8	8	8	8	7	5	7	7	7	7	6

G gene, *O* outcome

References

1. Tsao DA, Chang HJ, Hsiung SK, Huang SE, Chang MS, Chiu HH, Chen YF, Cheng TL, Shiu-Ru L (2010) Gene expression profiles for predicting the efficacy of the anticancer drug 5-fluorouracil in breast cancer. DNA Cell Biol 29:285–293
2. Latan MS, Laddha NC, Latani J, Imran MJ, Begum R, Misra A (2012) Suppression of cytokine gene expression and improved therapeutic efficacy of microemulsion- based tacrolimus cream for atopic dermatitis. Drug Deliv Transl Res 2:129–141
3. Albertin PS (1999) Longitudinal data analysis (repeated measures) in clinical trials. Stat Med 18:2863–2870
4. Yang X, Shen Q, Xu H, Shoptaw S (2007) Functional regression analysis using an F test for ongitudinal data with large numbers of repeated measures. Stat Med 26:1552–1566
5. Sverdlov L (2001) The fastclus procedure as an effective way to analyze clinical data. In: SUGI proceedings 26, paper 224, Long Beach, CA
6. Barthelemew DJ (1995) Spearman and the origin and development of factor analysis. Br J Math Stat Psychol 48:211–220
7. Anonymous (2012) SPSS Statistical Software version 18.0. Module dimension reduction, factor analysis, Online Help. www.spss.com. Accessed 29 Apr 2012
8. Wold H (1966) Estimation of principle components and related models by iterative least squares. In: Krishnaiah PR (ed) Multivariate analysis. Academic, New York, pp 391–420
9. Anonymous (2012) Factor analysis. Wikipedia, the free encyclopedia. Accessed 25 Apr 2012
10. Anonymous (2012) Partial least squares regression. http://en.wikipedia.org/wiki/Partial_least_squares_regression. Accessed 13 Apr 2012

11. Tenenhaus M, Vinzi VE, Chatelin YM, Lauro C (2005) PLS path modeling. Comput Stat Data Anal 48:159–205
12. Meng J (2012) Uncover cooperative gene regulations by microRNAs and transcription factors in glioblastoma using a nonnegative hybrid factor. www.cmsworldwide.com?ICASS2011. Accessed 29 Apr 2012
13. Hochreiter S, Clevert DA, Obermayer K (2006) A new summarization method for affymetrix probe level data. Bioinformatics 22:943–949
14. Bastien T (2004) PLS-Cox model: application to gene expression. COMPSTAT 2004; section: Partial Least Squares
15. Li X, Gill R, Cooper NG, Yoo JK, Datta S (2011) Modeling microRNA-mRNA interactions using pls regression in human colon cancer. BMC Med Genomics 4:44
16. Anonymous R (2012) Statistical software. Partial least squares, a free statistics and forecasting software. www.wessa.net/rwasp. Accessed 25 May 2012
17. Guinot C, Latreille J, Tenehaus M (2001) PLS path modeling and multiple table analysis. Application to cosmetic habits of women in Ile de France. Chem Intell Lab Syst 58:247–259

Chapter 17
Discriminant Analysis for Supervised Data

1 Summary

1.1 Background

MANOVA (multivariate analysis of variance) is used for the analysis of clinical trials with multiple outcome variables. However, its performance is poor if the relationship between the outcome variables is positive. Discriminant analysis is not affected by this mechanism.

1.2 Objective

To compare the performance of the two methods

1.3 Methods

A simulated case-study of three treatment regimens for the treatment of sepsis with multi-organ failure was used. The laboratory values after treatment were used as measure for treatment success.

1.4 Results

MANOVA with the treatment modality as predictor and the laboratory values as multiple outcome variables was not significant with a Pillai's test with $p=0.082$. Discriminant analysis of the three treatment modalities was significant at $p=0.049$.

T.J. Cleophas and A.H. Zwinderman, *Machine Learning in Medicine*,
DOI 10.1007/978-94-007-5824-7_17, © Springer Science+Business Media Dordrecht 2013

Discriminant analysis of the three subgroups analysis with Bonferroni adjustment showed that particularly the difference between two of the treatments was significant at p=0.017.

1.5   Conclusions

1. MANOVA (multivariate analysis of variance) is suitable for analysis of multiple outcome variables, but suffers from power loss, as the correlation between the outcome variables is, commonly, positive.
2. Discriminant analysis is more sensitive for testing studies with multiple outcome variables than the traditional MANOVA.
3. With multiple treatment modalities discriminant subgroup analyses with or without Bonferroni adjustment are possible to find out where the differences between the treatment groups are.
4. Discriminant analysis is available in SPSS statistical software and a wonderful and welcome methodology for the analysis of clinical trials with multiple outcome variables.

2   Introduction

Novel medical treatments are assessed in controlled clinical trials with health recovery as outcome. Usually, health recovery is measured as a simple outcome measure like the normalization of body temperature, or erythrocyte sedimentation rate. However, health recovery is not a simple entity, but a rather complex one with a multifactorial nature. And current clinical trials increasingly include multiple outcome variables like a battery of physical, laboratory and imaging tests. As the evaluation of novel treatments lies at the heart of health research for the benefit of mankind, statistical models that can handle multiple outcome variables are required.

Traditionally, MANOVA (multivariate analysis of variance) is used for the analysis of multiple outcome variables. However, a problem is that its performance is largely dependent on the levels of correlation between the outcome variables [1]. If negative, it will perform best, if positive it will do so worst. Unfortunately, in practice mostly strong positive correlations are present.

Discriminant analysis, although described as a method for binary data as early as 1935 by Sir Ronald Fisher [2], professor of statistics Cambridge UK, was proposed in its current version by William Klecka [3], professor of behavioral sciences, University of Cincinnati USA 1973. It eliminates the effects of the correlations between multiple outcome variables from the analysis by a technique called orthogonal linear modeling. It is currently widely used in the field of behavioral scien ces like social sciences, marketing, operational research and applied sciences [4], but is virtually unused in medicine, despite the omnipresence of multiple outcome variables in this branch of science. When searching Medline, we only found two genetic studies [5, 6], four diagnostic studies [7–10] and two clinical treatment trials [11, 12].

The current chapter using a simulated case-study assesses the performance of discriminant analysis versus traditional analysis in a clinical trial comparing the effect of different treatment modalities on a battery of laboratory outcome variables. This is an introduction to discriminant analysis, and was written as a hand-hold presentation accessible to clinicians, and as a must-read publication for those new to the methods. It is the author's experience, as a master class professor, that students are eager to master adequate command of statistical software. For their benefit all of the steps of the novel method from logging in to the final result using SPSS statistical software [13] will be given.

3   Some Theory

The underneath functions describe the two latent variables from Fig. 17.1 with 0.87, 0.90 etc. as best fit multiple linear regression coefficients.

$$\text{Liver function} = 0.87 \times \text{ASAT} + 0.90 \times \text{ALAT} + 0.85 \times \text{GammaGT}$$
$$+ 0.10 \times \text{creatinine} + 0.15 \times \text{creatininine clearance} - 0.05 \times \text{ureum}$$
$$\text{Renal function} = -0.10 \times \text{ASAT} - 0.05 \times \text{ALAT} + 0.05 \times \text{GammaGT}$$
$$+ 0.91 \times \text{creatinine} + 0.88 \times \text{creatininine clearance} + 0.90 \times \text{ureum}$$

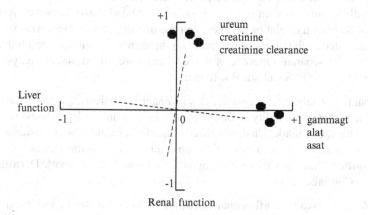

Fig. 17.1 Example of the relationships of six original variables with two latent variables, the liver and renal function. It can be observed that ureum, creatinine and creatinine clearance have a very strong positive correlation with renal function and the other three with liver function. It can be demonstrated that by slightly rotating both x and y-axes the model can be fitted even better. The orthogonal approach eliminates the effect of correlations between outcome variables on the final results when the latent variables are used for that purpose

Using the above functions we can now calculate the function-values for each patient, and subsequently use them as two independent outcome variables instead of the original six very much related outcome variables.

So far discriminant analysis is identical to factor analysis. However, discriminant analysis goes one step further. It includes a grouping predictor variable in the statistical model, e.g. treatment modality. The scientific question "is the treatment modality a significant predictor of a clinical improvement" is, subsequently, assessed by the question "is the outcome clinical improvement a significant predictor of the odds of having had a particular treatment". This reasoning may seem incorrect, using an outcome for making predictions, but, mathematically, it is no problem. It is just a matter of linear cause-effect relationships, but just the other way around, and it works very conveniently with "messy" outcome variables like in the example given.

4   Case-Study

The data of the case-study is given in the Appendix. Three treatment regimens for the treatment of sepsis with multi-organ failure were compared. The laboratory values after treatment were used as measure for treatment success.

MANOVA with the treatment modality as predictor and the laboratory values as multiple outcome variables was not significant with a Pillai's test with $p = 0.082$. Roy's largest root test was statistically significant, but did not meet its assumption of adequate power and accurate F approximation, and, so, the MANOVA concluded that there was not a significant difference between the three treatments. When performing multiple ANOVAs (analyses of variance) with a single lab value as outcome and the three treatment modalities as predictors, several (six out of nine) were statistically significant with p-values between 0.049 and 0.010. However, ANOVA does not account the relationship between the dependent variables. Also we were more interested in the effect of treatment on the combined outcome result than the effects on the separate outcome variables. Therefore, discriminant analysis was performed using SPSS statistical software.

Command: Analyze....Classify....Discriminant Analysis....enter Grouping Variable: treatment modality....Define Range 1–3....enter Independents: variables 1–10....Statistics: mark Unstandardized....mark Separate-groups covariance.... Continue....Classification: mark All groups equal....mark Summary table....mark Within-groups....mark Combined groups....Continue....Save: mark Discriminant scores....Continue....OK.

Table 17.1 gives two orthogonal discriminant functions (latent variables) of the ten outcome variables.

Function 1 = −3.062 + 0.000 gammagt + 0.001 asat + 0.002 alat − 0.006 bili +
        0.055 ureum + 0.004 creatinine − 0.023 creatinine clearance +
        0.001 esr − 0.015 c − reactive protein − 0.015 leucos.

4 Case-Study219

Table 17.1 The b-values (regression coefficients) of the functions

	Function	
	1	2
gammagt	0.000	–,007
asat	,001	,002
alat	,002	,003
bili	–,006	,008
ureum	,055	,019
creatinine	,004	–,003
creatinine clearance	–,023	,005
esr	,001	,005
c-reactive protein	–,024	,026
leucos	–,015	–,057
(Constant)	–3,062	,405

Unstandardized coefficients

Table 17.2 Test statistic of functions (otherwise called variates, or latent variables)

Test of function(s)	Wilks' Lambda	Chi-square	df	Sig.
1	,432	31,518	20	,049
2	,865	5,432	9	,795

Table 17.3 The mean function score for each treatment group

	Function	
Treatment modality	1	2
1,00	–,541	,392
2,00	–,641	–,508
3,00	1,604	–,028

Functions 2 can be described similarly using the results of function 2 of Table 17.1. The two functions are two best fit multiple linear regression models to summarize the outcome data.

Table 17.2 shows that according to function 1 there is a significant difference between the three treatment modalities at $p=0.049$, according to function 2 the differences are non significant at $p=0.795$. Table 17.3 gives the mean scores for each treatment group. Figure 17.2 gives treatment-group plots, the treatment-group centroids (dark squares) are the mean function scores for each group. The circles are the individual scores. The scores of function 1 is estimated along the x-axis, function 2 along the y-axis. The differences in mean scores along the x-axis is significantly different according to Wilk's test (Table 17.2). However with 3 groups we are unable to tell whether treatment 3 is different from 1, from 2, or from both 1 and 2. The difference in means scores along the y-axis is small and according to Wilk's test insignificant.

In order to test where the differences between the treatment groups are, 3 subgroup analyses are required.

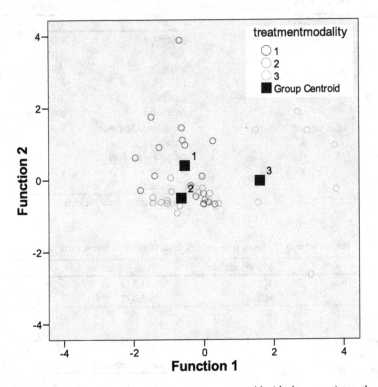

Fig. 17.2 Treatment-group plots, the treatment-group centroids (*dark squares*) are the mean function scores for each group. The *circles* are the individual scores. The scores in function 1 is estimated along the x-axis, function 2 along the y-axis. The between which of the groups the difference is, 3 subgroup analyses are required differences in mean scores along the x-axis is significantly different according to Wilk's test (above Table). However with 3 groups we are unable to tell whether 3 is different from 1, from 2, or from both 1 and 2. The difference in means sores along the y-axis is small and according to Wilk's test insignificant. In order to test

Command: Analyze....Classify....Discriminant Analysis....enter Grouping Variable: treatment modality....Define Range 2–3....enter Independents: variables 1–10.... Statistics: mark Unstandardized....mark Separate-groups covariance....Continue.... Classification: mark All groups equal....mark Summary table....mark Within-groups....mark Combined groups....Continue....Save: mark Discriminant scores.... Continue....OK.

The same procedure must be performed for comparing treatments 2 versus 1, and 1 versus 3.

Subgroup discriminant analyses of respectively treatment 2 versus 3, treatment 2 versus 1, and treatment 1 versus 3 are in Table 17.4.

Bonferroni adjustment of the rejection p-value with three tests equals

$$\text{p-value} = 0.05 * 2 / (3(3-1)) = 0.017 \ (* = \text{symbol of multiplication}).$$

Table 17.4 Subgroup discriminant analyses of respectively treatment 2 versus 3, treatment 2 versus 1, and treatment 1 versus 3

Test of function(s)	Wilks' Lambda	Chi-square	df	Sig.
1	,436	21,586	10	,017
Test of function(s)	Wilks' Lambda	Chi-square	df	Sig.
1	,446	15,340	10	,120
Test of function(s)	Wilks' Lambda	Chi-square	df	Sig.
1	,500	16,645	10	,083

Discriminant analysis, thus, shows that, even after correction for multiple testing, treatment 2 performs significantly better than treatment 3. The mean difference between treatment 1 and treatment 3 is virtually similar in magnitude, but probably due to the small sample size (treatment 2, n = 19; treatment 1, n = 14) it did not obtain statistical significance.

5 Discussion

In this case-study, in spite of a non significant MANOVA, a significant discriminant analysis between different treatment modalities was observed even after adjustment for multiple testing. Discriminant analysis may be more sensitive for testing studies with multiple outcome variables than the traditional MANOVA. Why so? This is probably because the functions produced by the orthogonal linear modeling may better fit multivariate data. Also a problem with MANOVA is that its performance is largely dependent on the correlations between the dependent variables. If negative it performs best, if positive it performs worst. In discriminant analysis the effect of the correlations between the dependent variables is eliminated.

Discriminant analysis is very similar to factor analysis (principle components analysis). Both apply orthogonal modeling in order to eliminate the effect of correlations between variables, but the former does so using the dependent variables, while the latter using the independent variables. Also, discriminant analysis goes one step further. It includes a grouping predictor variable in the statistical model, e.g. treatment modality. The scientific question "is the treatment modality a significant predictor of a clinical improvement" is, subsequently, assessed by the question "is the outcome clinical improvement a significant predictor of the odds of having had a particular treatment". This reasoning may seem incorrect, using an outcome for making predictions, but, mathematically, it is no problem. It is just a matter of linear cause-effect relationships, but just the other way around, and it works very conveniently with "messy" outcome variables like in the example given.

Both methods are currently often listed as machine learning methods because unlike traditional statistical tests they are able to handle large numbers of variables. They are modern computer intensive methods, sometimes also listed as artificial intelligence methods that allow computers to develop algorithms useful for the benefit of mankind and based on empirical data (the training data).

Because of the presence of a grouping predictor variable, discriminant analysis is sometimes called supervised machine learning, whereas factor analysis is called unsupervised learning.

These methods are also able to account many limitations of the traditional methods like accounting the relative importance of the multiple variables, their interactions and differences in units. And, in addition, as observed in the present report they may be more sensitive than traditional statistical methods.

Although developed and currently widely used in the fields of behavioral sciences, social sciences, marketing, operational research and applied sciences, these methods are virtually unused in clinical research [2–12].

We have to add that discriminant analysis is like (M)ANOVA based on linear regression, and most of its prior assumptions are similar, i.e., Gaussian data, homogeneity of variances, random data. However, a strong relationships between variables, sometimes called collinearity, is prohibitive for linear regression, but for discriminant analysis and many other machine learning methods it is not, either because they are not based on correlations but distances between individuals (cluster analysis) [14, 15] or because the effects of correlations is minimized by orthogonal modeling of variables (discriminant analysis, factor analysis, principal components analysis) [16].

6   Conclusions

1. Current clinical trials increasingly include multiple outcome variables.
2. MANOVA (multivariate analysis of variance) is suitable for analysis, but suffers from power loss as the correlation between the outcome variables is commonly positive.
3. Just like factor analysis (used for predictor variables), discriminant analysis (used for outcome variables) identifies 2 or more latent variables to explain all of the observed variables. Because both methods linearly models multiple variables along an x and y-axis, intervariable relationships do not further have to be taken into account.
4. Discriminant analysis goes one step further than factor analysis, and includes a grouping predictor variable, namely treatment modality, and answers the scientific question "is the outcome significantly related to the odds of having had a particular treatment".
5. Discriminant analysis is more sensitive for testing studies with multiple outcome variables than the traditional MANOVA.
6. With multiple treatment modalities subgroup analyses with or without Bonferroni adjustment are possible to find out where the differences between the treatment groups are.
7. Discriminant analysis is available in SPSS statistical software and a wonderful and welcome methodology for the analysis of clinical trials with multiple outcome variables.

7   Appendix: Data File

Var 1	var 2	var 3	var 4	var 5	var 6	var 7	var 8	var 9	var 10	var 11
24	12	26	4	5,7	83	−120	15	7	4	2
27	17	27	4	5,3	98	−101	14	4	3	2
23	15	28	4	5,4	83	−108	16	5	2	2
20	13	14	3	5,8	84	−132	13	6	7	2
17	13	13	3	5,4	88	−120	17	7	8	2
16	12	12	3	5,3	82	−132	14	6	9	3
15	17	15	2	6,4	89	−112	13	9	6	3
15	15	25	2	3,4	72	−125	15	4	6	3
14	14	24	2	2	78	−110	12	5	5	3
13	13	23	1	5,6	73	−102	16	7	4	3
18	19	19	1	6	79	−124	7	3	7	1
26	28	18	1	6,1	74	−112	8	4	8	1
37	17	17	3	5,4	78	−113	7	4	6	1
25	26	16	4	5,3	79	−112	16	5	4	1
36	15	29	2	6,3	89	−117	15	7	7	1
35	27	8	3	6	67	−107	17	7	6	1
24	14	27	4	4	76	−120	8	6	6	1
1320	1980	845	382	76	567	−7	123	7	26	3
1342	1870	756	203	45	578	−4	143	4	22	3
709	1980	645	254	42	865	−4	125	5	26	3
687	800	543	244	43	577	−6	143	6	23	3
25	29	15	3	45	566	−7	123	100	29	1
599	868	856	265	46	544	−9	165	120	30	1
900	759	856	287	45	532	−8	109	103	23	1
876	760	876	300	48	657	−5	102	90	24	1
945	879	946	266	87	685	−5	104	89	25	3
603	576	973	203	67	768	−5	104	79	24	3
1065	1487	865	265	63	675	−10	105	110	30	3
399	497	365	125	35	265	−25	76	47	18	1
25	16	29	201	38	345	−22	56	46	22	1
454	398	477	124	35	376	−21	65	45	21	1
497	355	476	136	25	333	27	45	36	20	1
477	389	365	154	24	334	−32	54	85	22	1
275	237	167	84	19	177	−56	43	28	16	2
266	263	187	54	14	198	−45	42	25	16	2
245	296	287	60	12	154	−54	48	24	15	2
290	188	165	67	14	156	−53	47	26	14	2
275	182	167	69	14	187	−57	45	27	15	2
267	139	177	74	13	167	−48	37	28	14	2
257	149	123	73	16	154	−45	36	25	16	2
276	200	154	72	15	243	−55	38	24	16	2
276	230	156	79	17	235	−54	34	23	15	2

(continued)

(continued)

Var 1	var 2	var 3	var 4	var 5	var 6	var 7	var 8	var 9	var 10	var 11
269	264	267	69	19	244	−55	45	29	15	1
162	276	287	87	16	175	−36	46	26	16	1
263	287	234	85	16	165	−37	47	27	16	1

Var 1 – gammagt (Var = variable), Var 2 – asat, Var 3 – alat ,Var 4 – bili, Var – 5 ureum, Var 6 – creatinine, Var – 7 creatinine clearance, Var 8 – esr (erythrocyte sedimentation rate), Var 9 – c-reactive protein, Var 10 – leucos, Var 11 – treatment modality

References

1. Cole DA, Maxwell SE, Arvey R, Salas E (1994) How the power of MANOVA can both increase and decrease as a function of the intercorrelations among the dependent variables. Psychol Bull 115:465–474
2. Fisher RA. The design of experiments. McMillan, London (First edition 1935; Ninth edition 1971)
3. Klecka WR (1973) Discriminant analysis. Google books, 1973, Google, Accessed 08 Jan 2012
4. Anonymous. Discriminant analysis. www.wikidepia.org/wiki/Discrimination_function_analysis
5. Ogah DM, Momoh OM, Dim NI (2011) Application of canonical discriminant analysis for assessment of genetic variation in Muscovy duck ecotypes in Nigeria. Egypt Poult Sci 31:429–436
6. Guo Y, Hastle T, Tibshirani R (2005) Regularized discriminant analysis and its application in microarrays. Biostatistics 1:1–18
7. Wernecke KD (2007) Clustering with parameters from blood tests. Wiley encyclopedia of clinical trials. Wiley, London
8. Wernecke KD (2007) Diagnosis of neuroborreliosis burgdorferi. Wiley encyclopedia of clinical trials. Wiley, London
9. Feigner JP, Sverdlov L (2001) The use of discriminant analysis to separate a study population by treatment subgroups in a clinical trial with a new pentapeptide antidepressant. J Appl Res 21:1–6
10. Adams KM (1979) Linear discriminant analysis in clinical neuropsychology. J Clin Neuropsychol 1:259–272
11. Glasson MJ, Stapleton F, Keay L, Sweeney D, Willcox MD (2003) Differences in clinical parameters and tear film of tolerant and intolerant contact lens wearers. Invest Ophthalmol Vis Sci 44:5116–5124
12. Fens N, Zwinderman AH, Van der Schee MP, De Nijs SB, Dijkers E, Roldaan AC, Cheung D, Bel EH, Sterk PJ (2009) Exhaled breath profiling enables discrimination of chronic obstructive pulmonary disease and asthma. Am J Respir Crit Care Med 180:1076–1082
13. SPSS Statistical Software. www.spss.com
14. Anonymous (2012) Cluster analysis. http://en.wikipedia.org/Clustert_analysis. Accessed 12 July 2012
15. Anonymous (2012) Sequence clustering. http://en.wikipedia.org/wiki/Sequence_analysis. Accessed 12 July 2012
16. Barthelemew DJ (1995) Spearman and the origin and development of factor analysis. Br J Math Stat Psychol 48:211–220

Chapter 18
Canonical Regression

1 Summary

1.1 Background

Canonical analysis assesses the combined effects of a set of predictor variables on a set of outcome variables, but is little used in clinical trials despite the omnipresence of multiple variables.

1.2 Objective

To assess the performance of canonical analysis as compared to traditional multivariate methods using MANCOVA (multivariate analysis of covariance).

1.3 Methods

As an example a simulated data file with 12 gene expression levels and 4 drug efficacy scores was used.

1.4 Result

The correlation coefficient between the 12 predictor and 4 outcome variables was 0.87 (p=0.0001) meaning that 76% of the variability in the outcome variables was explained by the 12 covariates. Repeated testing after the removal of 5 unimportant

T.J. Cleophas and A.H. Zwinderman, *Machine Learning in Medicine*,
DOI 10.1007/978-94-007-5824-7_18, © Springer Science+Business Media Dordrecht 2013

predictor and 1 outcome variable produced virtually the same overall result. MANCOVA identified identical unimportant variables, but was unable to provide overall statistics.

1.5   Conclusions

1. Canonical analysis is wonderful, because it can handle many more variables than traditional multivariate methods like MANCOVA.
2. At the same time it accounts for the relative importance of the separate variables, their interactions and differences in units.
3. Canonical analysis provides overall statistics of the effects of sets of variables, while traditional multivariate methods only provide the statistics of the separate variables.
4. Unlike other methods for combining the effects of multiple variables like factor analysis/partial least squares, canonical analysis is scientifically entirely rigorous.

We do hope that this chapter will stimulate clinical investigators to start using this wonderful method.

2   Introduction

In clinical trials the research question is often measured with multiple variables all of whom constitute a separate aspect or dimension of the question. For example, the expressions of a cluster of genes can be used as a functional unit to predict the efficacy of cytostatic treatment [1, 2], repeated measurements can be used as endpoint in randomized longitudinal trials [3, 4], and multi-item personal scores can be used for the evaluation of antidepressants [5]. Many more examples can be given.

ANOVA/ANCOVA (analysis of (co)variance) and MANOVA/MANCOVA (multivariate analysis of (co)variance) are the standard methods for the analysis of such data. A problem with these methods is, that they rapidly lose statistical power with increasing numbers of variables, and that computer commands may not be executed due to numerical problems with higher order calculations among components. Also, clinically, we are often more interested in the combined effects of the clusters of variables than in the separate effects of the different variables. As a simple solution composite variables can be used as add-up sums of separate variables, but add-up sums do not account the relative importance of the separate variables, their interactions, and differences in units. Canonical analysis can account all of that, and, unlike MANCOVA, gives, in addition to test statistics of the separate variables, overall test statistics of entire sets of variables.

Canonical analysis was invented by Harold Hotelling, a professor of statistics in the early 1970s at Chapel Hill University North Carolina, who used the term canonical stemming from Hebrew and meaning being an important part or element of something [6]. To date canonical analysis is little used in clinical trials despite the omnipresence of multiple variables. When searching Medline, we found apart from a few genetic trials [7, 8], and longitudinal trials [9, 10], virtually no clinical trials using it. Yet it is available in SPSS statistical software [11], although not in the Menu, but, rather, as one of the gems of SPSS, hidden in the Syntax program.

The current chapter was written to assess the performance of canonical analysis as compared to MANCOVA. As an example a simulated data file with 12 gene expression levels and four drug efficacy scores was used. We hope that this chapter will stimulate clinical investigators to start using this wonderful method.

3 Some Theory

Like ANCOVA/MANCOVA, canonical analysis is based on multiple linear regression, used to find the best fit correlation coefficients for your data. However, because it works with Wilks' statistic and beta distributions rather than Pillai's statistic and normal distributions, it is able to more easily calculate overall correlation coefficients between sets of variables. Yet it also assesses how a set of variables as a whole is related to its separate variables. In this way, an overall canonical model can be further improved by removing unimportant variables.

Canonical analysis is arithmetically equivalent to other methods for combining variables like factor-analysis/partial least squares analysis [12], but, conceptionally, it is very different. Unlike the latter, the former method does not produce new (latent) variables, but rather makes use of two sets of manifest variables. Also, unlike the latter, it complies with all of the requirements of traditional linear regression, and is, therefore, scientifically rigorous.

A canonical analysis should start with a correlation matrix. Variables with correlation coefficients larger than 0.80 are collinear and must be removed from the model. If in canonical models the clusters of predictor and outcome variables have a significant relationship, then this finding can, just like with linear regression, be used for making predictions in individual patients.

4 Example

A 250 patients' data-file (Appendix) was supposed to include 27 variables consistent of both patients' microarray gene expression levels and their drug efficacy scores. All variables were standardized by scoring them on 11 points linear scales (0–10). The following genes were highly expressed: the genes 1–4, 16–19, and 24–27. Four variables were supposed to represent drug efficacy scores and were clustered as the outcome variables 1–4.

For analyses SPSS statistical software [9] was used. The Menu does not offer canonical analysis, but the Syntax program does. In order to assess collinearity, a correlation matrix was first constructed.

Command: click File....click New....click Syntax....the Syntax Editor dialog box is displayed....enter the following text: "correlations/variables="and subsequently enter all of the gene-names....click Run.

A correlation matrix comes up (Table 18.1), and it shows that none of the correlation coefficients were larger than 0.8, and, so, high collinearity was not in the data. Then, MAN(C)OVA (multivariate analysis of (co)variance) was performed with the four drug efficacy scores as outcome variables and the 12 gene expression levels as covariates. We can now use the Menu command.

Command: click Analyze....click General Linear Model....click Multivariate.... Dependent Variables: enter the four drug efficacy scores....Covariates: enter the 12 genes....OK.

The Table 18.2 shows that MAN(C)OVA can be considered as another regression model with intercepts and regression coefficients. Just like AN(C)OVA it is based on normal distributions, and the results as given indicate that the model is adequate for the data. Generally, Pillai's method gives the best robustness. We can conclude that the genes 3, 16, 17, 19, 24, and 27 are significant predictors of all four drug efficacy outcome scores. Unlike AN(C)OVA, MAN(C)OVA does not give overall p-values, but rather separate p-values for separate covariates. However, in the given example the genes are considered a cluster of genes forming a single functional unit. Also the outcome variables are considered a cluster presenting different dimensions or aspects of drug efficacy. And, so, we are, particularly, interested in the combined effect of the set of covariates on the set of outcomes, rather than we are in modeling the separate variables. In order to asses the overall effect of the cluster of genes on the cluster of drug efficacy scores canonical regression is performed.

Command: click File....click New....click Syntax....the Syntax Editor dialog box is displayed....enter the following text: "manova" and subsequently enter all of the outcome variables....enter the text "WITH"....then enter all of the gene-names....then enter the following text: /discrim all alpha(1)/print=sig(eigen dim).... click Run.

Table 18.3, upper row, shows the result of the statistical analysis. The correlation coefficient between the 12 predictor and 4 outcome variables equals 0.87252. A squared correlation coefficient of 0.7613 means that 76 % of the variability in the outcome variables is explained by the 12 covariates. The cluster of predictors is a very significant predictor of the cluster of outcomes, and can be used for making predictions about individual patients with similar gene profiles. Repeated testing after the removal of separate variables gives us an idea about the relatively unimportant contributors as estimated by their coefficients, which are kind of canonical

Table 18.1 Correlation matrix of covariates

Correlations

		geneone	genetwo	gene-three	gene-four	gene-sixteen	gene-seventeen	gene-eighteen	gene-nineteen	gene-twentyfour	gene-twentyfive	genet-wentysix	gene-twentyseven
geneone	Pearson correlation	1	,644	,543	,504	,327	,349	,316	,396	,349	,343	,336	,283
	Sig. (2-tailed)		,000	,000	,000	,000	,000	,000	,000	,000	,000	,000	,000
	N	250	250	250	250	250	250	250	250	250	250	250	250
genetwo	Pearson correlation	,644	1	,776	,449	,554	,437	,418	,539	,497	,359	,509	,326
	Sig. (2-tailed)	,000		,000	,000	,000	,000	,000	,000	,000	,000	,000	,000
	N	250	250	250	250	250	250	250	250	250	250	250	250
genethree	Pearson correlation	,543	,776	1	,462	,579	,491	,477	,590	,524	,375	,521	,288
	Sig. (2-tailed)	,000	,000		,000	,000	,000	,000	,000	,000	,000	,000	,000
	N	250	250	250	250	250	250	250	250	250	250	250	250
genefour	Pearson correlation	,504	,449	,462	1	,384	,384	,420	,329	,285	,337	,243	,375
	Sig. (2-tailed)	,000	,000	,000		,000	,000	,000	,000	,000	,000	,000	,000
	N	250	250	250	250	250	250	250	250	250	250	250	250
genesix-teen	Pearson correlation	,327	,554	,579	,384	1	,634	,460	,636	,567	,336	,580	,284
	Sig. (2-tailed)	,000	,000	,000	,000		,000	,000	,000	,000	,000	,000	,000
	N	250	250	250	250	250	250	250	250	250	250	250	250
geneseven-teen	Pearson correlation	,349	,437	,491	,384	,634	1	,629	,496	,543	,343	,497	,431
	Sig. (2-tailed)	,000	,000	,000	,000	,000		,000	,000	,000	,000	,000	,000
	N	250	250	250	250	250	250	250	250	250	250	250	250

(continued)

Table 18.1 (continued)

Correlations

		geneone	genetwo	gene-three	gene-four	gene-sixteen	gene-seventeen	gene-eighteen	gene-nineteen	gene-twentyfour	gene-twentyfive	genet-wentysix	gene-twentyseven
geneeigh-teen	Pearson correlation	,316	,418	,477	,420	,460	,629	1	,504	,442	,278	,416	,415
	Sig. (2-tailed)	,000	,000	,000	,000	,000	,000		,000	,000	,000	,000	,000
	N	250	250	250	250	250	250	250	250	250	250	250	250
genenine-teen	Pearson correlation	,396	,539	,590	,329	,636	,496	,504	1	,565	,359	,584	,345
	Sig. (2-tailed)	,000	,000	,000	,000	,000	,000	,000		,000	,000	,000	,000
	N	250	250	250	250	250	250	250	250	250	250	250	250
genetwenty-four	Pearson correlation	,349	,497	,524	,285	,567	,543	,442	,565	1	,490	,732	,532
	Sig. (2-tailed)	,000	,000	,000	,000	,000	,000	,000	,000		,000	,000	,000
	N	250	250	250	250	250	250	250	250	250	250	250	250
genetwenty-five	Pearson correlation	,343	,359	,375	,337	,336	,343	,278	,359	,490	1	,512	,445
	Sig. (2-tailed)	,000	,000	,000	,000	,000	,000	,000	,000	,000		,000	,000
	N	250	250	250	250	250	250	250	250	250	250	250	250
genetwen-tysix	Pearson correlation	,336	,509	,521	,243	,580	,497	,416	,584	,732	,512	1	,477
	Sig. (2-tailed)	,000	,000	,000	,000	,000	,000	,000	,000	,000	,000		,000
	N	250	250	250	250	250	250	250	250	250	250	250	250
genetwenty-seven	Pearson correlation	,283	,326	,288	,375	,284	,431	,415	,345	,532	,445	,477	1
	Sig. (2-tailed)	,000	,000	,000	,000	,000	,000	,000	,000	,000	,000	,000	
	N	250	250	250	250	250	250	250	250	250	250	250	250

Table 18.2 Multivariate tests with the expression levels of 12 different genes with high expression as covariates and drug efficacy scores as outcome variables

	Effect value	F	Hypothesis df	Error df	p-value
Intercept	0.043	2.657	4.0	234.0	0.034
Gene 1	0.006	0.362	4.0	234.0	0.835
Gene 2	0.27	1.595	4.0	234.0	0.176
Gene 3	0.042	2.584	4.0	234.0	0.038
Gene 4	0.013	0.744	4.0	234.0	0.563
Gene 16	0.109	7.192	4.0	234.0	0.0001
Gene 17	0.080	5.118	4.0	234.0	0.001
Gene 18	0.23	1.393	4.0	234.0	0.237
Gene 19	0.092	5.938	4.0	234.0	0.0001
Gene 24	0.045	2.745	4.0	234.0	0.029
Gene 25	0.017	1.037	4.0	234.0	0.389
Gene 26	0.027	1.602	4.0	234.0	0.174
Gene 27	0.045	2.751	4.0	234.0	0.029

Four statistical tests were performed, but they consistently produced four identical p-values. We only report Pillai's test

F Fisher statistic, df degree of freedom

Table 18.3 Canonical regression results

Numbers variables (covariates v outcome variables)

	Canon cor	Sq cor	Wilks L	F	Hypoth df	Error df	p
12 v 4	0.87252	0.7613	0.19968	9.7773	48.0	903.4	0.0001
7 v 4	0.87054	0.7578	0.21776	16.227	28.0	863.2	0.0001
7 v 3	0.87009	0.7571	0.22043	22.767	21.0	689.0	0.0001

cor correlation coefficient, sq squared, L lambda, $hypoth$ hypothesis, df degree of freedom, p p-value, v versus

b-values (regression coefficients). The larger they are, the more important they are. Table 18.4 left column gives an overview. The outcome 3, and the genes 2, 4, 18 and 25 contributed little to the overall result. When restricting the model by removing the variables with canonical coefficients smaller than 0.05 or larger than −0.05 (the middle and right columns of Table 18.4), the results were largely unchanged. And so were the results of the overall tests (Table 18.3, 2nd and 3rd rows). Seven versus three variables produced virtually the same correlation coefficient but with much more power (lambda increased from 0.1997 to 0.2204, the F value from 9.7773 to 22.767, in spite of a considerable fall in the degrees of freedom (Table 18.3)). It, therefore, does make sense to try and remove the weaker variables from the model ultimately to be used.

The weakest contributing covariates of the MANCOVA (Table 18.2) were virtually identical to the weakest canonical predictors (Table 18.4), suggesting that the two methods are closely related and one method confirms the results of the other.

Table 18.4 Raw and standardized (z transformed) canonical coefficients, otherwise called canonical weights (the multiple b-values of canonical regression)

	12 v 4	7 v 4	7 v 3
Raw			
Outcome 1	−0.24620	−0.24603	0.25007
Outcome 2	−0.20355	−0.19683	0.20679
Outcome 3	−0.02113	−0.02532	
Outcome 4	−0.07993	−0.08448	0.09037
Gene 1	0.01177		
Gene 2	−0.01727		
Gene 3	−0.05964	−0.08344	0.08489
Gene 4	−0.02865		
Gene 16	−0.14094	−0.13883	0.13755
Gene 17	−0.12897	−0.14950	0.14845
Gene 18	−0.03276		
Gene 19	−0.10626	−0.11342	0.11296
Gene 24	−0.07148	−0.07024	0.07145
Gene 25	−0.00164		
Gene 26	−0.05443	−0.05326	0.05354
Gene 27	0.05589	0.04506	−0.04527
Standardized			
Outcome 1	−0.49754	−0.49720	0.50535
Outcome 2	−0.40093	−0.38771	0.40731
Outcome 3	−0.03970	−0.04758	
Outcome 4	−0.15649	−0.16539	0.17693
Gene 1	0.02003		
Gene 2	−0.03211		
Gene 3	−0.10663	−0.14919	0.15179
Gene 4	−0.04363		
Gene 16	−0.30371	−0.29918	0.29642
Gene 17	−0.23337	−0.27053	0.26862
Gene 18	−0.06872		
Gene 19	−0.23696	−0.25294	0.25189
Gene 24	−0.18627	−0.18302	0.18618
Gene 25	−0.00335		
Gene 26	−0.14503	−0.14191	0.14267
Gene 27	0.12711	0.10248	−0.10229

5   Discussion

Canonical analysis is wonderful, because it can handle many more variables than
    MAN(C)OVA, and, at the same time account for the relative importance of the separate variables, their interactions and differences in units. The current chapter includes only 27 variables, but Waaijenberg and Zwinderman [8] showed that studies including several thousands of variables are possible, provided that variable reduction without too much loss of information is performed. The elastic net and lasso method are computationally intensive methods available for that purpose [13, 14].

The current chapter shows that canonical analysis with 12 predictor and 4 outcome variables produces a very significant result, and that in the example given we can predict the combined outcome from the combined set of predictors with 76% certainty. A limitation of canonical analysis is that often weak canonical correlations with little practical value are involved. For example, R (= canonical correlation coefficient)=0.3 means that the squared $R=0.3^2=0.09=9\%$, and that only 9% certainty is given. Also, compared to factor-analysis/partial least squares, it can handle only two sets of variables. SPSS in the Syntax program can only handle linear relationships, but non-linear canonical regression is a possibility, and is offered by SPSS through the add-on module Categories.

Canonical analysis may be arithmetically equivalent to factor-analysis/partial least squares analysis [8], it is, conceptionally, very different. Unlike factor-analysis/partial least squares, it does not produce new variables, but, rather, uses two sets of manifest variables. It leads to the conclusion, that a set of variables is, or is not, an important predictor to a set of outcome variables. The latter method produces entirely new variables, otherwise called latent variables, which are basically hypothetical and do not really exist. We should admit that factor-analysis/partial least square analysis is more flexible, because it can be used to produce *multiple* latent predictor and outcome variables. However it does not comply with the traditional requirements of linear regression like normal distributions, adequate sample sizes, adjustments for multiple testing etc., and is sometimes called soft modeling. Also, empirical, instead of parametric confidence intervals, are applied, with hypothesis testing based on bootstrap re-sampling.

In contrast, canonical analysis is entirely rigorous. It uses the same assumptions as linear regression, including the assumption that the predictor variables should not highly correlate with one another. Therefore, a canonical analysis should always start with a correlation matrix: variables with correlation coefficients larger than 0.80 are collinear and must be removed from the model. With factor-analysis/partial least squares collinear variables is no problem. They are, simply, included in the latent variables after orthogonal linear weighting.

Canonical analysis is a rigorous method, but little is known about the effects of violating the linear regression assumptions. Also the sample size issue is not entirely solved: Stevens [15] recommends that there should be at least 20 times as many cases as variables for reliable estimates. Outliers can greatly reduce the magnitude of the correlation coefficients, and traditional scatter plots are helpful for that purpose.

6 Conclusions

1. Canonical analysis is wonderful, because it can handle many more variables than traditional multivariate methods like MAN(C)OVA can.
2. At the same time it accounts for the relative importance of the separate variables, their interactions and differences in units.

3. Canonical analysis provides overall statistics of the effects of sets of variables, while traditional multivariate methods only provide the statistics of the separate variables.
4. Unlike other methods for combining the effects of multiple variables like factor analysis/partial least squares, canonical analysis is scientifically entirely rigorous.

7   Appendix: Datafile of the Example Used in the Present Chapter

G1	G2	G3	G4	G16	G17	G18	G19	G24	G25	G26	G27	O1	O2	O3	O4
8	8	9	5	7	10	5	6	9	9	6	6	6	7	6	7
9	9	10	9	8	8	7	8	8	9	8	8	8	7	8	7
9	8	8	8	8	9	7	8	9	8	9	9	9	8	8	8
8	9	8	9	6	7	6	4	6	6	5	5	7	7	7	6
10	10	8	10	9	10	10	8	8	9	9	9	8	8	8	7
7	8	8	8	8	7	6	5	7	8	8	7	7	6	6	7
5	5	5	5	5	6	4	5	5	6	6	5	6	5	6	4
9	9	9	9	8	8	8	8	9	8	3	8	8	8	8	8
9	8	9	8	9	8	7	7	7	7	5	8	8	7	6	6
10	10	10	10	10	10	10	10	10	8	8	10	10	10	9	10
2	2	8	5	7	8	8	8	9	3	9	8	7	7	7	6
7	8	8	7	8	6	6	7	8	8	8	7	8	7	8	8
8	9	9	8	10	8	8	7	8	8	9	9	7	7	8	8
7	7	8	8	8	9	10	7	9	4	8	8	9	8	7	7
3	4	3	8	4	4	4	3	4	3	4	4	4	4	3	4
7	8	8	5	8	8	7	6	7	7	8	7	10	8	8	7
8	8	8	8	6	8	5	1	9	7	7	8	7	7	8	6
7	8	8	8	8	9	8	7	10	10	9	8	9	9	9	9
8	4	3	8	3	5	5	3	2	10	1	0	5	3	4	3
8	7	6	10	8	8	7	6	4	4	5	5	7	7	7	5
9	9	10	8	8	9	7	7	8	9	8	9	8	7	8	7
6	6	6	6	4	5	4	5	3	9	3	4	4	5	4	3
8	8	8	7	7	7	8	6	8	7	9	4	6	7	8	9
9	9	10	9	10	10	7	10	10	10	10	10	8	8	8	5
8	7	8	8	9	8	9	8	8	8	8	8	8	8	8	9
8	5	5	4	2	1	1	0	0	1	0	0	3	2	4	5
6	6	6	6	5	6	3	5	4	4	4	5	5	6	3	4
7	8	9	8	8	9	9	6	9	8	8	10	9	8	7	7
8	8	8	7	7	7	7	6	7	8	7	8	7	6	6	6
8	8	8	8	9	8	9	8	9	8	9	9	9	8	7	8
7	7	7	6	7	7	9	7	7	7	7	8	8	6	7	7
9	9	9	9	6	9	8	7	8	8	8	9	8	8	8	8
10	10	10	10	9	9	10	5	10	2	9	9	8	10	8	8
9	8	9	9	8	7	7	8	9	9	9	9	8	5	9	7
8	9	9	9	8	7	7	6	7	8	8	8	8	7	8	6

(continued)

(continued)

G1	G2	G3	G4	G16	G17	G18	G19	G24	G25	G26	G27	O1	O2	O3	O4
3	4	2	5	4	2	2	4	4	4	3	4	6	2	3	2
8	8	9	9	8	8	8	8	8	8	8	8	8	8	8	8
8	6	7	6	7	7	8	6	7	6	5	5	6	7	7	6
10	10	10	10	7	10	10	8	10	10	10	10	10	8	8	7
8	10	9	8	8	8	7	6	7	7	10	8	9	8	8	7
8	8	8	8	8	8	7	8	7	8	8	8	9	9	8	7
5	7	7	8	5	7	7	3	1	6	3	10	5	6	6	5
10	9	9	10	7	9	9	9	9	9	9	8	8	9	7	7
9	7	7	9	3	6	4	2	1	8	2	1	6	6	6	6
8	8	10	8	9	8	7	8	8	7	8	8	9	6	5	7
6	8	8	8	9	10	10	9	10	9	9	10	9	8	5	5
8	8	8	8	10	8	7	10	8	8	7	10	9	7	8	6
6	5	5	6	6	6	4	6	3	5	0	3	7	5	5	3
9	9	9	8	8	9	8	7	6	7	8	10	8	8	8	6
9	10	8	8	9	10	10	9	7	8	9	7	8	8	7	7
8	8	8	9	6	8	7	6	8	9	8	8	7	7	6	5
8	5	6	7	8	8	7	7	4	6	7	6	8	8	7	6
4	1	4	9	0	0	7	0	0	10	0	10	0	0	0	0
5	5	7	5	7	7	8	5	7	7	5	5	7	7	7	7
5	5	6	5	4	4	4	3	3	2	3	3	3	4	3	3
7	9	9	10	5	9	9	9	9	6	7	6	10	7	10	9
10	10	10	10	8	9	9	6	7	8	8	10	7	7	7	6
8	8	8	8	6	9	8	7	7	6	6	2	7	7	7	5
6	6	7	9	8	8	7	6	1	9	0	4	6	7	7	6
6	7	7	7	6	5	5	5	5	7	3	5	7	6	6	8
9	9	9	9	8	8	9	6	8	7	6	10	8	7	7	8
7	7	7	7	6	8	8	6	7	7	7	8	6	6	5	10
9	7	8	9	8	10	8	9	8	9	7	9	7	7	8	3
8	9	9	8	7	8	7	8	8	6	7	8	7	8	7	6
8	8	8	8	6	8	8	5	8	9	8	7	7	7	6	5
7	7	7	7	4	5	6	6	3	6	7	7	1	5	6	5
9	10	9	9	8	9	8	8	9	8	9	9	8	7	8	8
8	9	9	8	8	8	8	7	8	7	8	9	6	6	5	6
7	8	8	8	6	7	7	6	8	5	7	7	7	6	7	5
4	2	2	6	5	5	4	4	6	4	3	2	4	6	7	2
5	5	7	5	5	5	5	2	2	9	5	5	4	5	5	4
9	9	10	9	7	8	7	8	8	9	8	8	8	8	6	9
8	8	8	8	7	7	7	9	8	9	7	8	7	7	5	6
8	8	9	8	8	9	5	9	8	5	7	6	8	8	8	6
9	9	9	9	6	8	8	4	7	5	6	6	7	7	8	8
9	8	8	8	7	9	9	9	10	10	10	10	10	9	7	10
9	9	9	8	8	8	8	7	7	7	7	7	8	8	8	7
8	5	7	9	2	8	8	2	9	10	1	9	5	5	5	5
7	6	9	8	5	7	7	6	5	7	4	4	6	7	6	7
8	8	9	8	6	7	7	6	8	7	7	10	8	7	8	6
10	10	10	10	8	10	10	7	8	8	7	8	9	9	9	7
9	9	6	6	4	5	5	5	2	3	5	4	2	3	3	3

(continued)

(continued)

G1	G2	G3	G4	G16	G17	G18	G19	G24	G25	G26	G27	O1	O2	O3	O4
3	3	3	8	0	7	0	0	0	7	0	10	0	0	8	8
5	4	4	7	4	4	4	2	0	4	2	8	3	3	3	3
8	10	10	10	7	8	7	10	10	9	8	10	10	9	9	8
5	8	8	8	7	8	8	6	7	7	7	10	7	8	6	6
7	4	5	9	5	8	7	5	5	8	0	7	6	6	6	6
5	6	5	8	10	9	0	8	8	8	8	5	8	8	5	4
7	5	7	6	3	6	6	3	5	6	6	5	5	5	5	5
10	8	9	8	8	8	8	6	8	8	6	6	8	7	5	8
10	10	10	10	10	10	8	10	9	10	10	10	10	9	9	8
6	6	4	5	0	5	5	5	5	8	5	9	6	4	5	5
10	3	7	9	0	5	7	7	10	8	10	10	5	5	5	5
5	7	8	7	8	7	8	7	8	6	7	6	8	6	7	6
9	10	9	9	10	6	6	7	9	8	8	8	10	7	7	10
10	10	10	10	9	10	10	9	10	10	10	10	9	9	9	9
10	10	10	10	10	10	10	8	10	10	6	10	7	6	8	8
7	7	7	8	7	8	8	6	8	8	7	7	8	7	8	8
9	5	7	9	6	8	8	4	6	7	4	5	6	5	5	4
9	9	10	8	8	9	8	7	8	8	7	9	8	7	5	7
6	6	5	4	4	4	4	3	4	3	4	5	4	5	5	5
7	8	8	9	7	5	4	7	10	8	8	8	6	4	4	7
8	6	6	5	7	6	0	8	7	9	7	7	7	7	6	7
6	8	8	9	9	9	9	5	9	8	7	9	9	5	5	9
9	5	6	7	10	10	8	7	8	9	10	10	8	8	7	8
8	7	8	5	8	7	4	5	8	5	5	9	3	5	3	5
7	8	7	4	8	8	8	7	7	6	6	7	8	7	7	7
8	7	10	10	10	10	10	10	10	10	10	10	9	6	4	8
5	9	10	5	9	9	6	8	10	10	10	9	8	7	9	9
9	6	6	7	10	10	6	6	9	10	10	9	10	10	10	9
0	4	7	5	10	8	9	9	9	7	8	7	8	9	9	9
4	8	8	6	9	9	7	2	9	9	9	9	8	8	8	7
8	8	10	8	7	7	5	5	5	10	8	3	7	7	6	7
9	10	10	7	5	4	0	7	10	10	10	10	5	4	5	9
10	10	10	10	7	0	0	8	2	8	1	0	4	5	3	3
10	8	8	8	5	5	8	8	10	10	10	10	6	6	5	5
7	10	10	8	10	10	8	8	10	9	10	10	7	8	10	6
10	9	9	6	9	9	0	9	10	8	9	9	8	7	10	7
8	10	8	5	7	6	5	7	10	10	10	10	6	6	7	7
10	8	8	7	8	8	7	5	10	8	8	10	8	8	7	8
8	7	8	8	10	10	2	1	8	10	8	8	9	7	9	10
8	8	8	8	6	7	7	4	8	8	7	7	7	5	6	7
7	9	8	8	9	8	8	7	9	9	9	7	10	9	7	7
8	8	9	9	7	7	8	7	7	8	7	7	7	7	8	8
8	7	8	7	8	8	8	7	8	8	7	8	8	8	8	7
8	7	7	8	7	7	8	7	8	8	7	8	7	7	8	7
8	8	8	8	7	6	8	6	9	8	7	9	8	8	6	6
8	8	8	9	9	6	8	9	8	9	10	10	8	8	8	5

(continued)

(continued)

G1	G2	G3	G4	G16	G17	G18	G19	G24	G25	G26	G27	O1	O2	O3	O4
7	8	8	6	8	9	9	6	8	8	8	8	8	8	6	8
7	9	9	8	6	8	8	5	8	7	5	9	7	5	7	4
10	10	10	8	9	8	8	8	10	10	10	10	10	10	9	9
6	8	7	8	9	8	10	8	8	9	9	8	8	7	7	5
8	8	8	8	8	8	8	8	8	8	5	10	8	8	8	7
10	0	0	10	0	7	5	0	0	3	0	10	0	0	0	0
8	5	9	4	6	8	8	5	6	6	4	5	6	5	5	4
9	9	9	9	8	8	8	7	7	3	0	9	7	7	8	8
8	9	8	8	8	8	8	8	8	9	9	8	8	8	9	5
7	7	7	7	7	7	7	5	7	7	7	5	8	7	5	6
9	9	9	9	7	7	8	8	8	7	8	6	8	6	6	7
5	7	4	10	0	10	10	0	5	5	0	10	0	0	0	0
9	9	9	9	9	10	10	9	10	10	10	10	10	10	5	5
8	8	9	7	7	8	8	7	8	7	7	8	8	8	6	8
9	10	10	7	9	9	8	4	9	9	9	8	8	7	9	9
10	10	10	10	10	10	9	7	10	10	10	9	7	7	5	9
8	6	9	9	7	9	8	5	6	6	5	5	6	7	5	4
7	7	8	5	8	8	7	6	5	5	7	4	5	6	6	6
9	10	10	10	9	8	9	8	8	8	8	9	9	8	6	7
7	7	6	6	4	6	6	4	4	6	3	5	4	4	4	4
8	8	8	8	9	8	7	9	10	3	7	10	9	8	7	7
8	8	8	8	7	8	5	8	10	10	7	10	8	7	7	7
10	10	10	10	10	10	10	10	10	10	10	9	10	10	10	10
10	10	10	10	9	10	10	9	10	10	10	10	9	9	9	8
9	10	10	10	8	10	10	8	10	10	10	10	9	8	8	7
4	6	8	8	7	7	7	5	4	7	5	9	6	6	7	5
8	8	8	7	7	8	9	7	7	5	7	4	8	9	9	9
8	8	8	8	6	7	7	4	6	10	6	6	7	7	7	5
8	8	4	8	5	5	5	1	0	5	0	10	2	2	2	2
7	7	7	7	7	8	8	4	7	7	6	6	6	6	6	6
8	7	7	8	10	9	8	9	10	9	8	9	9	8	7	8
9	9	7	8	9	8	8	8	8	8	9	8	9	7	8	6
5	3	4	3	4	5	3	5	2	3	5	4	4	2	4	7
6	8	8	8	9	9	8	7	9	8	9	10	8	8	7	7
9	10	10	10	6	8	9	8	0	10	10	10	10	9	6	9
4	5	5	7	4	4	5	4	2	4	2	7	5	5	3	3
8	8	8	8	10	10	10	10	10	10	10	8	10	7	7	7
9	9	9	9	10	8	8	8	8	8	7	8	9	9	8	8
10	10	10	10	8	8	8	8	8	8	8	9	9	9	8	8
10	10	10	9	10	10	10	10	10	10	10	10	10	10	10	10
10	10	10	10	7	5	5	5	6	8	8	5	8	5	5	10
7	8	8	8	4	5	5	4	5	4	5	8	7	6	8	4
8	8	8	8	5	8	8	5	5	5	5	7	6	6	5	5
8	6	8	5	5	5	5	3	3	9	3	2	5	3	5	3
10	10	10	10	10	10	10	10	9	10	10	10	10	9	10	10

(continued)

(continued)

G1	G2	G3	G4	G16	G17	G18	G19	G24	G25	G26	G27	O1	O2	O3	O4
7	7	7	7	7	8	8	5	6	7	7	9	6	7	5	5
8	7	7	8	8	9	5	5	6	7	6	5	7	7	6	6
10	10	10	10	9	10	10	10	9	10	10	10	10	9	10	5
7	9	9	9	8	9	8	8	9	8	8	7	9	10	8	8
9	8	8	8	9	9	8	7	10	8	9	10	9	8	7	8
8	6	6	7	5	7	5	4	5	2	5	5	6	5	5	4
8	9	9	9	6	8	7	6	6	5	5	7	7	6	7	6
7	8	9	9	9	10	10	7	10	5	8	8	10	10	5	9
9	8	8	8	8	9	7	8	0	5	7	10	8	8	9	2
10	10	10	10	6	10	7	8	10	9	2	8	9	9	7	6
10	10	9	10	10	10	10	9	10	10	10	10	10	10	9	10
8	9	9	8	8	8	8	8	8	8	8	8	8	8	9	8
8	10	10	10	8	8	8	8	9	9	9	8	9	8	9	8
8	8	8	5	5	8	8	8	6	8	10	5	7	7	5	7
6	6	7	7	6	7	5	2	5	5	5	0	6	10	6	6
10	10	10	10	5	10	10	10	10	10	10	10	10	10	5	10
8	7	8	8	7	9	9	7	6	8	8	8	7	7	5	6
8	7	8	7	8	8	8	8	9	9	8	9	8	7	7	6
7	7	7	8	8	9	8	7	8	8	8	9	7	7	7	7
10	10	10	10	10	10	10	10	10	10	10	10	10	10	10	10
10	10	10	9	7	9	9	7	8	8	8	7	8	8	8	8
10	10	10	10	10	10	10	5	10	10	10	10	9	10	9	9
10	10	10	10	10	10	10	10	10	10	10	10	10	10	9	9
10	10	10	9	10	10	9	9	10	6	10	10	10	10	7	9
7	9	9	8	9	10	9	8	8	8	8	8	8	7	5	7
9	9	9	9	9	9	8	8	9	9	8	7	9	8	8	8
6	5	5	7	1	5	6	5	5	10	5	10	3	0	5	5
10	10	10	10	7	10	10	10	10	10	10	10	10	10	5	10
8	9	10	9	9	10	9	9	9	10	10	9	10	9	10	9
6	8	8	9	3	8	5	5	5	5	7	6	5	5	6	6
9	9	9	9	5	8	5	6	9	9	8	10	8	8	8	8
8	9	9	8	5	8	8	8	8	8	7	9	7	7	5	7
6	7	7	7	6	6	6	3	3	6	0	6	5	5	5	5
8	8	8	9	7	8	8	8	5	8	7	10	7	7	7	6
8	8	9	6	6	7	5	5	10	5	0	10	7	7	5	5
8	9	9	7	6	7	7	6	9	7	7	7	7	6	7	7
8	4	6	7	3	6	6	6	0	6	0	9	6	5	4	6
9	9	9	9	9	8	8	8	7	8	8	8	8	8	8	8
6	7	7	6	6	6	4	4	5	6	8	5	2	3	3	4
6	7	7	7	4	6	4	4	4	8	4	5	6	7	7	5
8	7	7	9	7	10	5	6	8	8	6	9	6	7	6	7
10	10	10	9	8	7	8	7	8	9	9	8	5	5	5	4
8	7	8	10	8	9	6	7	8	7	8	8	8	8	7	8
8	9	7	8	9	8	8	7	8	7	5	9	6	8	8	8
7	7	5	7	8	8	6	6	9	7	8	8	7	7	6	7

(continued)

(continued)

G1	G2	G3	G4	G16	G17	G18	G19	G24	G25	G26	G27	O1	O2	O3	O4
9	9	10	8	8	8	6	5	10	10	10	10	7	7	5	6
8	6	9	9	8	9	8	9	8	7	7	8	9	9	7	8
7	7	8	9	7	7	7	8	7	8	9	7	6	8	7	7
7	7	8	7	8	7	8	7	8	8	6	5	7	8	7	7
9	10	9	9	8	7	9	9	6	6	6	6	7	9	8	8
7	7	7	6	6	6	9	9	8	3	5	8	6	9	9	8
9	10	7	8	7	5	10	10	10	10	7	10	6	8	9	7
4	6	5	7	4	4	4	3	10	9	10	9	4	4	3	4
8	8	8	8	8	8	7	6	10	8	10	10	10	8	8	7
8	6	3	8	6	8	5	1	10	7	10	10	7	7	8	6
7	7	8	7	8	8	8	6	4	6	5	9	7	7	5	6
9	10	9	9	8	7	7	9	7	4	7	4	5	9	8	6
10	10	8	9	7	6	5	8	6	7	6	6	7	9	9	5
4	6	9	8	9	9	7	7	0	1	0	0	8	7	9	4
8	8	7	7	4	6	5	5	4	4	4	5	3	5	4	3
8	8	8	8	7	7	8	6	8	7	8	9	6	7	7	9
8	8	3	4	9	8	7	6	6	8	6	8	7	8	8	5
6	8	8	7	5	7	7	7	8	7	8	8	7	7	7	7
9	10	9	8	7	8	8	8	6	7	6	8	8	8	8	8
4	5	5	7	6	9	6	8	9	9	8	8	8	10	10	7
6	8	8	8	8	8	8	6	5	4	6	6	9	9	9	8
9	9	7	7	4	6	5	3	3	5	3	6	4	3	7	5
8	9	5	5	7	8	7	6	6	7	6	6	7	6	7	4
10	10	8	7	7	8	6	5	6	8	5	7	5	6	8	8
7	9	8	7	7	8	10	8	9	8	7	8	9	8	8	7
9	9	9	9	6	3	4	4	7	8	6	8	6	3	4	4
6	5	5	7	6	7	6	4	9	8	8	9	7	7	7	6
9	9	9	9	9	10	10	8	7	6	8	8	8	8	8	7
8	9	9	7	9	9	8	8	3	5	3	6	7	6	7	8
8	8	8	8	4	5	5	8	8	2	8	7	8	7	8	6
8	9	9	4	8	9	9	7	6	7	8	6	9	7	8	7
10	9	7	7	7	8	8	8	8	7	5	7	7	7	7	6

G gene, *O* outcome

References

1. Tsao DA, Chang HJ, Hsiung SK, Huang SE, Chang MS, Chiu HH, Chen YF, Cheng TL, Shiu-Ru L (2010) Gene expression profiles for predicting the efficacy of the anticancer drug 5-fluorouracil in breast cancer. DNA Cell Biol 29:285–293
2. Latan MS, Laddha NC, Latani J, Imran MJ, Begum R, Misra A (2012) Suppression of cytokine gene expression and improved therapeutic efficacy of microemulsion- based tacrolimus cream for atopic dermatitis. Drug Deliv Transl Res 2:129–141
3. Albertin PS (1999) Longitudinal data analysis (repeated measures) in clinical trials. Stat Med 18:2863–2870
4. Yang X, Shen Q, Xu H, Shoptaw S (2007) Functional regression analysis using an F test for longitudinal data with large numbers of repeated measures. Stat Med 26:1552–1566

5. Sverdlov L (2001) The fastclus procedure as an effective way to analyze clinical data. In: SUGI proceedings 26, paper 224, Long Beach, CA
6. Anderson TW (1990) Two of Harold Hotelling's contributions to multivariate analysis. Technical report no. 40. National Science Foundation Grant DMS 89–04851
7. Naylor MG, Lin X, Weiss ST, Raby BA, Lange C (2010) Using canonical correlation analysis to discover genetic regulatory variants. PLoS One 5(5):e10395
8. Waaijenberg S, Zwinderman AH (2007) Penalized canonical correlation analysis to quantify the association between gene expression and DNA markers. BMC Proc 1(Suppl 1):S122–S125
9. Alonso A, Geys H, Molenberghs G, Kenward MG, Vangeneugden T (2004) Validation of surrogate markers in multiple randomized clinical trails with repeated measurements, Canonical correlation approach. Biometrics 60:845–853
10. Meredith W, Tisak J (1982) Canonical analysis of longitudinal and repeated measures data with stationary weights. Psychometrika 47:47–67
11. SPSS statistical software. www.spss.com. 12 June 2012
12. Sun L, Ji S, Yu S, Ye J (2009) On the equivalence between canonical correlation analysis and orthonormalized partial least squares. In: Proceedings of conference on artificial intelligence 2009, AAAI (Association Advancement Artificial Intelligence) Publications, Palo Alto, CA, pp 1230–1235
13. Zou H, Hastie T (2005) Regularization and variable selection via the elastic net. J Roy Stat Soc B 67:301–320
14. Tibshirani R (1996) Regression shrinkage and selection via the lasso. J Roy Stat Soc B 58:267–288
15. Stevens J (1986) Applied multivariate statistics for social sciences. Erlbaum, Hillsdale

Chapter 19
Fuzzy Modeling

1 Summary

1.1 Background

Fuzzy logic can handle questions to which the answers may be "yes" at one time and "no" at the other, or may be partially true and untrue. Pharmacodynamic data deal with questions like "does a patient respond to a particular drug dose or not", or "does a drug cause the same effects at the same time in the same subject or not". Such questions are typically of a fuzzy nature, and might, therefore, benefit from an analysis based on fuzzy logic.

1.2 Objective

This chapter assesses whether fuzzy logic can improve the precision of predictive models for pharmacodynamic data.

1.3 Methods and Results

1. The quantal pharmacodynamic effects of different induction dosages of thiopental on numbers of responding subjects was used as the first example. Regression analysis of the fuzzy-modeled outcome data on the imput data provided a much better fit than did the un-modeled output values with r-square values of 0.852 (F-value=40.34) and 0.555 (F-value 8.74) respectively.
2. The time-response effect propranolol on peripheral arterial flow was used as a second example. Regression analysis of the fuzzy-modeled outcome data on the

imput data provided a better fit than did the un-modeled output values with r-square values of 0.990 (F-value = 416) and 0.977 (F-value = 168) respectively.

1.4   Conclusions

We conclude that fuzzy modeling may better than conventional statistical methods fit and predict pharmacodynamic data, like, for example, quantal dose response and time response data. This may be relevant to future pharmacodynamic research.

2   Introduction

Lofti Zadeh, professor of science at Berkeley, published in 1964 the concept of fuzzy truths, as answers that may be "yes" at one time and "no" at the other, or that may be partially true and partially untrue [1]. He developed an analytical model based on this concept. When you think of real life, you can imagine many things that are not entirely certain, and it is remarkable, therefore, that it took over 20 years before this analytical model became successfully implemented in science [1]. Nowadays Tokyo subway traffic uses fuzzy logic running and braking systems [2], and Maserati sportscars have a fuzzy logic automatic transmission with one position for forward instead of the usual three or four, and with much better performance.

In the field of medicine fuzzy logic is little used in spite of the, typically, uncertain character of this branch of science. When searching for published papers we found a few papers on diagnostic imaging [3] and clinical decision analysis [4–6]. In clinical pharmacology fuzzy logic has been applied for pharmacological treatment decision analyses [7–9], and structure-activity modeling [10]. However, we found no papers on fuzzy logic and pharmacodynamic modeling. Often the basic molecular mode of action of a drug is unknown, and pharmacodynamics is, then, used as a surrogate for studying the pharmacological response to a drug of the body. By its very nature pharmacodynamics can be argued to be particularly fuzzy. For example, the answer to the question "does a patient respond or not to a particular thiopental induction dose", or questions like "does propranolol cause the same effects at the same time in the same subject or not" are typically questions of a fuzzy nature, and might, thus, benefit from an analysis based on fuzzy logic.

In the present chapter we study whether fuzzy logic can improve the precision of predictive models for pharmacodynamic data, i.e., models that better fit the observed data, and, thus, better predict future data. We hope that the examples given will stimulate researchers analyzing pharmacodynamic data to more often apply fuzzy methodologies.

3 Some Fuzzy Terminology

3.1 Fuzzy Memberships

The universal spaces are divided into equally sized parts called membership functions

3.2 Fuzzy Plots

Graphs summarizing the fuzzy memberships of (for example) the imput values (Fig. 19.2 upper graph).

3.3 Linguistic Membership Names

Each fuzzy membership is given a name, otherwise called linguistic term.

3.4 Linguistic Rules

The relationships between the fuzzy memberships of the imput data and those of the output data (the method of calculation is shown in the underneath examples).

3.5 Triangular Fuzzy Sets

A common way of drawing the membership function with on the x-axis the imput values, on the y-axis the membership grade for each imput value.

3.6 Universal Space

Defined range of imput values, defined range of output values.

4 First Example, Dose-Response Effects of Thiopental on Numbers of Responders

We will use as an example the quantal pharmacodynamic effects of different induction dosages of thiopental on numbers of responding subjects (Table 19.1, left two columns). It is usually not possible to know what type of statistical distribution the experiment is likely to follow, sometimes Gaussian, sometimes very skewed. A pleasant aspect of fuzzy modeling is that it can be applied with any type of statistical distribution and that it is particularly suitable for uncommon and unexpected non linear relationships. Quantal response data are often presented in the literature as S-shape dose-cumulative response curves with the dose plotted on a logarithmic scale, where the log transformation has an empirical basis. We will, therefore, use a logarithmic regression model. SPSS Statistical Software 17.0 [11] is used for analysis.

Command: Analyze...regression...curve estimation...dependent variable: data second column...independent variable: data first column...logarithmic...ok.

The analysis produces a moderate fit of the data (Fig. 19.1 upper curve) with an r-square value of 0.555 (F-value 8.74, p-value 0.024).
We, subsequently, fuzzy-model the imput and output relationships (Fig. 19.2).
First of all, we create linguistic rules for the imput and output data.
For that purpose we divide the universal space of the imput variable into fuzzy memberships with linguistic membership names:

imput-*zero, -small, -medium, -big, -superbig*.

Then we do the same for the output variable:

output-*zero, -small, -medium, -big*.

Subsequently, we create linguistic rules.
Figure 19.2 shows that imput-*zero* consists of the values 1 and 1.5.

The value 1 (100% membership) has 4 as outcome value (100% membership of output-*zero*).
The value 1.5 (50% membership) has 5 as outcome value (75% membership of output-*zero*, 25% of output-*small*).

The imput-*zero* produces $100\% \times 100\% + 50\% \times 75\% = 137.5\%$ membership to output-*zero*, and $50\% \times 25\% = 12.5\%$ membership to output-*small*, and so, output-zero is the most important output contributor here, and we forget about the small contribution of output-*small*.
Imput-*small* is more complex, it consists of the values 1.5, and 2.0, and 2.5.

The value 1.5 (50% membership) has 5 as outcome value (75% membership of output-*zero*, 25% membership of output-*small*).
The value 2.0 (100% membership) has 6 as outcome value (50% membership of outcome-*zero*, and 50% membership of output-*small*).
The value 2.5 (50% membership) has 9 as outcome value (75% membership of output-*small* and 25% of output-*medium*).

The imput-*small* produces $50\% \times 75\% + 100\% \times 50\% = 87.5\%$ membership to output-*zero*, $50\% \times 25\% + 100\% \times 50\% + 50\% \times 75\% = 100\%$ membership to output-*small*, and $50\% \times 25\% = 12.5\%$ membership to output-*medium*. And so, the output-*small* is the most important contributor here, and we forget about the other two.
For the other imput memberships similar linguistic rules are determined:

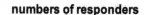

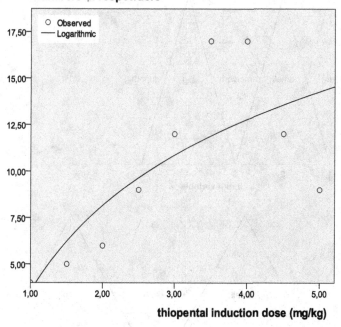

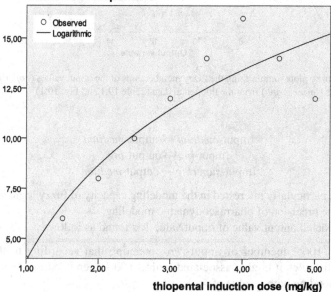

Fig. 19.1 Pharmacodynamic relationship between induction dose of thiopental (x-axis, mg/kg) and number of responders (y-axis). The un-modeled curve (*upper curve*) fits the data less well than does the modeled (*lower curve*) with r-square values of 0.555 (F-value = 8.74), and 0.852 (F-value = 40.34) respectively

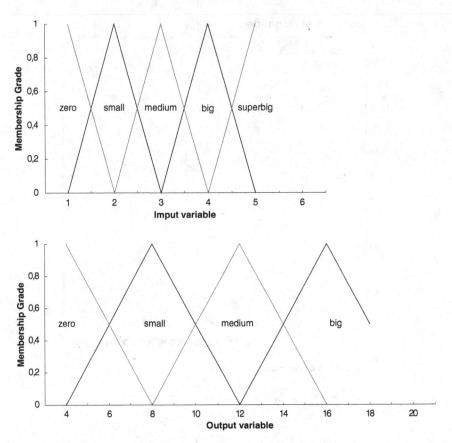

Fig. 19.2 Fuzzy plots summarizing the fuzzy memberships of the imput values (*upper graph*) and output values (*lower graph*) from the thiopental data (Table 19.1 and Fig. 19.1)

Imput-*medium* → output-*medium*

Imput-*big* → output-*big*

Imput-*superbig* → output-*medium*

We are, particularly interested in the modeling capacity of fuzzy logic in order to improve the precision of pharmacodynamic modeling.

The modeled output value of imput value 1 is found as follows.

Value 1 is 100% member of imput-*zero*, meaning that according to the above linguistic rules it is also associated with a 100% membership of output-*zero* corresponding with a value of 4.

Value 1.5 is 50% member of imput-*zero* and 50% imput-*small*. This means it is 50% associated with the output-*zero* and –*small* corresponding with values of $50\% \times (4 + 8) = 6$.

Table 19.1 Quantal pharmacodynamic effects of different induction dosages of thiopental on numbers of responding subjects

Imput values	Output values	Fuzzy-modeled output
Induction dosage of thiopental (mg/kg)	Numbers of responders (n)	Numbers of responders (n)
1	4	4
1.	5	5
2	6	8
2.5	9	10
3	12	12
3.5	17	14
4	17	16
4.5	12	14
5	9	12

For all of the imput values modeled output values can be found in this way. Table 19.1 right column shows the results. We perform a logarithmic regression on the fuzzy-modeled outcome data similar to that for the un-modeled output values. The fuzzy-modeld output data provided a much better fit than did the un-modeled output values (Fig. 19.2, lower curve) with an r-square value of 0.852 (F-value = 40.34) as compared to 0.555 (F-value 8.74) for the un-modeled output data.

5   Second Example, Time-Response Effect of Propranolol on Peripheral Arterial Flow

The pharmacodynamic effect of a single oral dose of 120 mg of propranolol on peripheral arterial is used as a second example (Table 19.2 left two columns). The magnitude of the pharmacodynamic response is estimated as absolute change of fore arm flow using a venous occlusion plethysmograph. Like with quantal dose response curves it is, usually, impossible to know what statistical distribution the curves are likely to follow. This is no problem for fuzzy modeling. But we use a quadratic regression model (second order model), because it is the simplest model after the linear model and fits many nonlinear data. SPSS Statistical Software 17.0 [11] is used for analysis.

Command: Analyze…regression…curve estimation…dependent variable: data second column…independent variable: data first column…quadratic…ok.

The analysis produces a good fit of the data (Fig. 19.3 upper graph) with an r-square value as large as 0.977 with an F-value of 168.

We, subsequently, fuzzy-model the imput and output relationships.

First of all, we create linguistic rules for the imput and output data.

Table 19.2 Time-response effect of single oral dose of 120 mg propranolol on peripheral arterial flow

Imput values	Output values	Fuzzy-modeled output
Hours after oral adminis-tration of 120 mg propranolol	Peripheral arterial flow (ml/100 ml tissue/min)	Peripheral arterial flow (ml/100 ml tissue/min)
1	20	20
2	12	14
3	9	8
4	6	6
5	5	4
6	4	4
7	5	4
8	6	6
9	9	8
10	12	14
11	20	20

For that purpose we divide the universal space of the imput variable into fuzzy memberships with linguistic membership names:

imput-*null, -zero, -small, -medium, -big, -superbig.*
Then we do the same for the output variable:

output-*zero, -small, -medium, -big, -superbig.*

Subsequently, we will create linguistic rules.
Imput-*null* consists of the values 1 and 2 (Fig. 19.4).

The value 1 (100% membership) has 20 as outcome value (100% membership of output-*superbig*)
The value 2 (50% membership) has 12 as outcome value (100% membership of output-*medium*).

The imput-*null* produces $100\% \times 100\% = 100\%$ membership to output-superbig, $50\% \times 100\% = 50\%$ membership to output-*medium*. And so, output-*superbig* is the most important contributor here, we forget about the other one.
Imput-*zero* consists of the values 2, 3, 4.

The value 2 (50% membership) has 12 as outcome value (100% membership of output-*medium*).
The value 3 (100% membership) has 9 as outcome value (75% membership of outcome-*small*, and 25% membership of output-*medium*).
The value 4 (50% membership) has 6 as outcome value (50% membership of output-small and 50% of output-*zero*).

The imput-*zero* produces $50\% \times 100\% + 100\% \times 25\% = 125\%$ membership to output-*medium*, $100\% \times 75\% + 50\% \times 50\% = 100\%$ membership to output-*small*, and $50\% \times 50\% = 25\%$ membership to output-*zero*. And so, output-*medium* is the most important contributor here, and we forget about the other two.

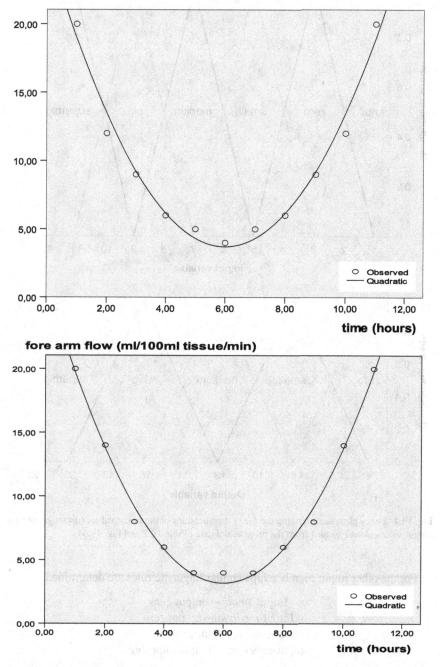

Fig. 19.3 Pharmacodynamic relationship between the time after oral administration of 120 mg of propranolol (x-axis, hours) and absolute change in fore arm flow (y-axis, ml/100 ml tissue/min). The un-modeled curve (*upper curve*) fits the data slightly less well than does the modeled (*lower curve*) with r-square values of 0.977 (F-value = 168), and 0.990 (F-value = 416) respectively

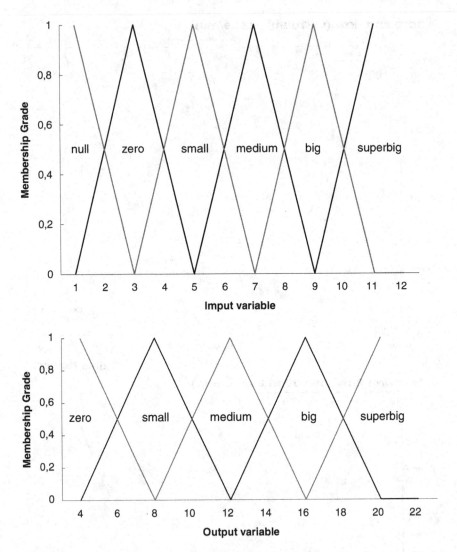

Fig. 19.4 Fuzzy plots summarizing the fuzzy memberships of the imput values (*upper graph*) and output values (*lower graph*) from the propranolol data (Table 19.2 and Fig. 19.3)

For the other imput memberships similar linguistic rules are determined:

Imput-*small* → output-*zero*
Imput-*medium* → output-*zero*
Imput-*big* → output-*small*
Imput-*superbig* → output-*superbig*

We are, particularly, interested in the modeling capacity of fuzzy logic in order to improve the precision of pharmacodynamic modeling.

The modeled output value of imput value 1 is found as follows.

Value 1 is 100% member of imput-*null*, meaning that according to the above linguistic rules it is also associated with a 100% membership of output-*superbig* corresponding with a value of 20.

Value 2 is 50% member of imput-*null* and 50% imput-*zero*. This means it is 50% associated with the output-*superbig* and –*small* corresponding with values of $50\% \times (8 + 20) = 14$.

For all of the imput values modeled output values can be found in this way. Table 19.2 right column shows the results. When performing a quadratic regression on the fuzzy-modeled outcome data similar to that shown above, the fuzzy-modeled output values provided a better fit than did the un-modeled output values (Fig. 19.3, upper an. lower curves) with r-square values of 0.990 (F-value = 416) and 0.977 (F-value = 168).

6   Discussion

Biological processes are full of variations. Statistical analyses do not provide certainties, only chances, particularly, the chances that prior hypotheses are true or untrue. Fuzzy statistics is different from conventional statistical methods, because it does not assess the chance of entire truths, but, rather, the presence of partial truths. The advantage of fuzzy statistics compared to conventional statistics is that it can answer questions to which the answers are "yes" and "no" at different times, or partly "yes" and "no" at the same time. Additional advantages are that it can be used to match any set of im- and output data, including incomplete and imprecise data, and nonlinear functions of unknown complexity as sometimes observed with pharmacodynamic data. The current paper suggests, indeed, that fuzzy logic may better fit and, thus, better predict pharmacodynamic data than conventional methods.

We have only shown the simplest method of fuzzy modeling with a single imput and a single output variable. Just like with multiple regression, multiple imput variables are possible, and are capable of adequately modeling complex chemical and engineering processes [12]. To date, complex fuzzy models are rarely applied in medicine, but one recent clinical study successfully used ten imput variables including sex, age, smoking, and clinical grade, to predict tumor relapse time [4]. The problem is that such calculations soon get very complex and can not be carried out on a pocket calculator like in our examples. Statistical software is required. Fuzzy logic is not yet widely available in statistical software programs, and it is not in SPSS [11] or SAS [13], but several user-friendly programs do exist [14–16].

We conclude.

1. Fuzzy logic is different from conventional statistical methods, because it does not asses the chance of entire truths but rather the presence of partial truths.
2. The advantage of fuzzy statistics compared to conventional statistics is that it can answer questions to which the answers are "yes" and "no" at different times, or partly "yes" and "no" at the same time.

3. Additional advantages are that it can be used to match any set of im- and output data, including incomplete and imprecise data, and nonlinear functions of unknown complexity as sometimes observed with pharmacodynamic data.
4. Fuzzy modeling may better than conventional statistical methods fit and predict quantal dose response and time response data.

We hope that the examples given will stimulate researchers analyzing pharmacodynamic data to more often apply fuzzy methodologies.

7   Conclusions

Fuzzy logic can handle questions to which the answers may be "yes" at one time and "no" at the other, or may be partially true and untrue. Pharmacodynamic data deal with questions like "does a patient respond to a particular drug dose or not", or "does a drug cause the same effects at the same time in the same subject or not". Such questions are typically of a fuzzy nature, and might, therefore, benefit from an analysis based on fuzzy logic. This chapter assesses whether fuzzy logic can improve the precision of predictive models for pharmacodynamic data.

1. The quantal pharmacodynamic effects of different induction dosages of thiopental on numbers of responding subjects was used as the first example. Regression analysis of the fuzzy-modeled outcome data on the imput data provided a much better fit than did the un-modeled output values with r-square values of 0.852 (F-value = 40.34) and 0.555 (F-value 8.74) respectively.
2. The time-response effect propranolol on peripheral arterial flow was used as a second example. Regression analysis of the fuzzy-modeled outcome data on the imput data provided a better fit than did the un-modeled output values with r-square values of 0.990 (F-value = 416) and 0.977 (F-value = 168) respectively.

We conclude that fuzzy modeling may better than conventional statistical methods fit and predict pharmacodynamic data, like, for example, quantal dose response and time response data. This may be relevant to future pharmacodynamic research.

References

1. Zadeh LA (1965) Fuzzy sets. Inf Control 8:338–353
2. Hirota K (1993) Subway control. In: Hirota (ed) Industrial applications of fuzzy technology. Springer, Tokyo, pp 263–269
3. Fournier C, Castelain B, Coche-Dequeant B, Rousseau J (2003) MRI definition of target volumes using logic metho for three-dimensional conformal radiation therapy. Int J Rad Oncol 55:225–233
4. Catto JW, Linckens DA, Abbod MF, Chen M, Burton JL, Feeley KM, Hamdy FC (2003) Artificial intelligence in predicting bladder cancer outcome: a comparison of fuzzy modelling and artificial intelligence. Clin Cancer Res 9:4172–4177

5. Bates JH, Young MP (2003) Applying fuzzy logic to medical decision making in the intensive care. Am J Resp Crit Care Med 167:948–952
6. Caudrelier JM, Vial S, Gibon D, Kulik C, Swedko P, Boxwala A (2004) An authoring tool for fuzzy logic based decision support systems. Medinfo 9:1874–1879
7. Naranjo CA, Bremmer KE, Bazoon M, Turksen IB (1997) Using fuzzy logic to predict response to citalopram in alcohol dependence. Clin Pharmacol Ther 62:209–224
8. Helgason CM, Jobe TH (2005) Fuzzy logic and continuous cellular automata in warfarin dosing of stroke patients. Curr Treat Options Cardiol Med 7:211–218
9. Helgason CM (2004) The application of fuzzy logic to the prescription of antithrombotic agents in the elderly. Drugs Aging 21:731–736
10. Russo M, Santagati NA (1998) Medicinal chemistry and fuzzy logic. Inf Sci 105:299–314
11. SPSS Statistical Software (2010) www.SPSS.com. Accessed 22 Oct 2010
12. Ross J (2004) Fuzzy logic with engineering applications, 2nd edn. Wiley, Chichester
13. SAS Statistical Software (2010) www.SAS.com. Accessed 22 Oct 2010
14. FuzzyLogic (2010) fuzzylogic.sourceforge.net. Accessed 22 Oct 2010
15. Fuzzy logic: flexible environment for exploring fuzzy systems (2010) www.wolfram.com. Accessed 22 Oct 2010
16. Defuzzification methods. www.mathworks.com. Accessed 22 Oct 2010

Chapter 20
Conclusions

1 Introduction

The current book is an introduction to the wonderful methods that statistical software offers in order to analyze large and complex data. A nice thing about the novel methodologies, is, that, unlike the traditional methods like analysis of variance (ANOVA) and multivariate analysis of variance (MANOVA), they can not only handle large data files with numerous exposure and outcome variables, but also can do so in a relatively unbiased way.

2 Limitations of Machine Learning

It is time that we addressed some limitations of the novel methods.

First, if you statistically test large data, you will almost certainly find some statistically significant effects. They may be statistically significant, but are very small, and, therefore, often clinically irrelevant.

Second, multiple variables require multiple testing, and raise the risk of significant effects due to chance, rather than a clinical mechanism.

Third, testing without a prior hypothesis raises the chance of type I errors of finding an effect which does not exist.

Fourth, large samples are at risk of being overpowered: less power would be more adequate to demonstrate a clinical effect of a desired magnitude.

3   Serendipities and Machine Learning

We should add that, in clinical research to date, important novel discoveries are generally not serendipities, otherwise called sensational and unexpected novelties [1]. The princes of Serendip went on random journeys, and always came home with sensational novelties. In clinical research serendipities have been rare so far. Maybe the medical use of penicilline and nitroglycerine were serendipities, but inventions in clinical research were mostly the result of hard work and prospective testing of prior hypotheses.

Also in the past, medical conclusions based on uncontrolled observations appeared to be subsequently frequently wrong. Indeed, the findings from randomized controlled trials may be more accurate and reliable than those from uncontrolled data files. But are they completely meaningless? First, data mining and other machine learning methods to establish a priori hypothesized health risks may be hard but are not impossible. Second, today computer files from clinical data are very large and may include hundreds of variables. Testing such data without a prior hypothesis is not essentially different from data dredging/ fishing, and your finding may be true in less than 10% of the cases, but 10% is better than 0%.

4   Machine Learning in Other Disciplines and in Medicine

Machine learning was highly efficacious in geoscience, marketing research, anthropology and other disciplines [2]. In medicine it is little used so far. Some examples are:

1. Bio-informatics and genetics research (DNA sequencing for disease susceptibilities) [3].
2. Data mining of clinical trials as an alternative to laborious meta-analyses [4].
3. Adverse drug surveillance [5].

The lack of use in medicine is probably a matter of time, now that many machine learning methods are available in SPSS and other statistical software.

5   Conclusions

We hope that the book is helpful to clinicians, medical students, and clinical investigators to improve their expertise in the field of machine learning, and that it facilitates the analysis of the large and complex data files widely available in the electronic health records of modern health facilities.

References

1. Clancy C, Simpson L (2002) Looking forward to impact: moving beyond serendipities. Health Serv Res 37:14–23
2. Anonymous. Data mining (2012) Wikipedia. http://en.wikipedia.org/wiki/Data_mining. Accessed 30 Aug 2012
3. Zhu X, Davidson I (2007) Knowledge discovery and data mining. Hershey, New York, pp 163–189
4. Zhu X, Davidson I (2007) Knowledge discovery and data mining. Hershey, New York, pp 31–48
5. Bate A, Lindquist M, Edwards I, Olsson S, Orre R, Lansner A, De Freitas R (1998) A Bayesian neural network method for adverse drug reaction signal generation. Eur J Clin Pharmacol 54:315–321

Index